Kluwer Academic Publishers

P.O. Box 17, 3300 AA Dordrecht, The Netherlands

W0227424

Dear Reader

We would very much appreciate receiving your suggestions and criticisms on the *Plant Tissue Culture Manual*. They will be most helpful during our preparations for future supplements.

Would you please answer the questions listed below, and send your comments with any further suggestions you may have, to *Ir. A. Plaizier* at the above-mentioned address.

Thank you for your assistance!

A. Plaizier
Publisher

——

PLANT TISSUE CULTURE MANUAL

1. What errors have you found? (list page numbers and describe mistakes)
2. What protocols do you find to be confusing or lacking in detail? (list chapter numbers and page numbers and describe problems)
3. What protocols do you feel should be replaced in future supplements with newer (better) methods?
4. What new topics or other material would you like to see included in future supplements?

Please print or type your answers in the space below, and continue overleaf.

Name: Date:

Address:

Plant Tissue Culture Manual
Supplement 6, February 1996

INSTRUCTIONS FOR SUPPLEMENT 6

Preliminary pages
Replace: Table of Contents

End matter
Replace: Index

PLANT TISSUE CULTURE MANUAL SUPPLEMENT 6

PLANT TISSUE CULTURE MANUAL
Supplement 6

Edited by:

K. LINDSEY
Department of Biological Sciences, University of Durham, U.K.

Kluwer Academic Publishers
Dordrecht / Boston / London

Library of Congress Cataloging-in-Publication Data

Plant tissue culture manual: fundamentals and applications / edited by K. Lindsey
   p.   cm.
  Includes bibliographical references and index.

  1. Plant tissue culture—Laboratory manuals.   I. Lindsey, K.
QK725.P587   1991   90–26765
581'.0724—dc20

Manual

ISBN-13: 978-0-7923-3874-1   e-ISBN-13: 978-94-009-0181-0
DOI: 10.1007/978-94-009-0181-0

Published by Kluwer Academic Publishers,
P.O. Box 17, 300 AA Dordrecht, The Netherlands.

Kluwer Academic Publishers incorporates
the publishing programmes of
D. Reidel, Martinus Nijhoff, Dr W. Junk and MTP Press.

Sold and distributed in the U.S.A. and Canada
by Kluwer Academic Publishers,
101 Philip Drive, Norwell, MA 02061, U.S.A.

In all other countries, sold and distributed
by Kluwer Academic Publishers Group,
P.O. Box 322, 3300 AH Dordrecht, The Netherlands.

Printed on acid-free paper

Contents

Preface

---

* Included in Supplement 6.

5. Restriction fragment analysis of somaclones
   R.H. Potter, As, Norway

6. Virus elimination and testing
   M.C. Coleman and W. Powell, Dundee, UK

7. Clonal propagation of *Citrus*
   T.S. Rangan, Pasadena, USA

8. Clonal propagation of eucalypts
   J.A. McComb, Perth, Australia

*9. Cryopreservation of plant tissue cultures: the example of embryogenic
   tissue cultures from conifers
   P.J. Charest, J. Bongs and K. Klimaszewska, Ontario and New
   Brunswick, Canada

SECTION D: DIRECT GENE TRANSFER & PROTOPLAST FUSION

1. Gene transfer by particle bombardment
   T.M. Klein, S. Knowlton and R. Arentzen, Wilmington, USA

2. Transformation of pollen by particle bombardment
   D. Twell, T.M. Klein and S. McCormick, Albany, USA

3. Electrical fusion of protoplasts
   M.G.K. Jones, Murdoch, Australia

4. Cybrid production and selection
   E. Galun and D. Aviv, Rehovot, Israel

5. Fluorescence-activated analysis and sorting of protoplasts and somatic
   hybrids
   D.W. Galbraith, Lincoln, USA

6. RFLP analysis of organellar genomes in somatic hybrids
   E. Pehu, Helsinki, Finland

7. Isolation and uptake of plant nuclei
   P.K. Saxena and J. King, Saskatoon, Canada

8. *In situ* hybridization to plant metaphase chromosomes: Radioactive and
   non-radioactive detection of repetitive and low copy number genes
   J. Veuskens, S. Hinnisdaels and A. Mouras, Talence, France

9. Chemical fusion of protoplasts
   P. Anthony, R. Marchant, N.W. Blackhall, J.B. Power, M.R. Davey,
   Nottingham, UK

---

* Included in Supplement 6.

SECTION E: REPRODUCTIVE TISSUES

1. *In vitro* fertilisation of maize
   E. Kranz, Hamburg, Germany

2. Endosperm culture
   S. Stirn and H.-J. Jacobsen, Bonn, Germany

3. Endosperm culture
   B.M. Johri and P.S. Srivastava, Delhi, India

4. Hybrid embryo rescue
   A. Agnihotri, New Delhi, India

5. *In vitro* culture of *Brassica juncea* zygotic proembryos
   Chun-ming Liu, Zhi-hong Xu and Nam-Hai Chua,
   Singapore and New York, USA

SECTION F: MUTANT SELECTION

1. Use of chemical and physical mutagens *in vitro*
   P.J. Dix, Maynooth, Ireland

*2. *In vitro* culture, mutant selection, genetic analysis and transformation of
   Physcomitrella patens
   David Cove, Leeds, UK

SECTION G: SECONDARY METABOLITES

1. Tropane alkaloid biosynthesis *in vitro*
   R. Robins, Colney, Norwich, UK

2. Anthocyanin biosynthesis *in vitro*
   A. Komamine and K. Kakegawa, Tsukuba City, Ibaraki, Japan

3. Biosynthesis of monoterpene indole alkaloids *in vitro*
   W.G.W. Kurz, K. Constabel, R. Tyler, Saskatchewan, Canada

SECTION H: TISSUE CULTURE TECHNIQUES FOR FUNDAMENTAL STUDIES

1. Establishment of photoautotrophic cell cultures
   W. Hüsemann, Münster, Germany

2. *Zinnia* mesophyll culture system to study xylogenesis
   M. Sugiyama and H. Fukuda, Sendai, Japan

---

* Included in Supplement 6.

---

* Included in Supplement 6.

Plant Tissue Culture Manual **B12**, 1–20. 1996.

Production of fertile transgenic wheat by microprojectile bombardment

DIRK BECKER & HORST LÖRZ

Institut für Allgemeine Botanik der Universität Hamburg (AMPII), Ohnhorststraße 18, 22609 Hamburg, Germany

Introduction

The development of plant transformation techniques during the past decade has made it possible to improve plants by introduction of cloned genes. For most dicotyledonous species, the *Agrobacterium*-mediated transformation system can be used to generate many transformants while for most of the monocotyledonous species, this transformation system still needs to be improved. However, in the last two years two reports were published describing the regeneration of transgenic fertile rice by using *Agrobacterium tumefaciens* as vector [4, 11]. Although this is a very recent development, the future prospects also for other cereals like maize, wheat and barley will surely be investigated in the near future. Of the various approaches to gene transfer, three transformation methods have led predominantly to the production of transgenic plants:
– protoplast based direct gene transfer,
– tissue electroporation,
– microprojectile-mediated gene transfer.
In the past considerable progress has been made in establishing efficient *in vitro* culture systems for most cereals. However, thus far embryogenic suspension cultures are the only reliable source for totipotent protoplasts. Nevertheless, it is very difficult and time-consuming to start and maintain these cultures. Furthermore, regeneration capability has been observed to decline gradually during cultivation in cereal suspension cultures. The direct DNA transfer into isolated protoplasts, induced by polyethylene glycol (PEG) or electric pulses is a successful and routinely used method to obtain transformed cell lines, but the regeneration of transgenic plants still remains difficult. Only in rice and maize it has been possible until now to obtain fertile, transgenic plants by protoplast transformation [21, 7, 8]. The fundamental problem of this transformation method is the continous loss of embryogenic capacity of the suspension cultures during long term culture [13], occurrence of somaclonal variation, and expenditure of labour and energy. As an alternative, microprojectile-mediated gene transfer [20] or tissue electroporation [6] have the potential to overcome these limitations. The essence of microprojectile systems for plant genetic transformation is to use high velocity particles to penetrate cell walls and to introduce DNA into intact cells thus circumventing the host range limitation of *Agrobacterium* and the problems of plant regeneration from protoplasts.

The transfer of DNA into cells and tissues with embryogenic capacity takes place with high efficiency. The choice of appropriate target cells is of major importance as there are only few tissues and cells capable of plant regeneration. Using embryogenic suspension cells and embryogenic callus cultures, successful transformation and regeneration of cereals, such as maize, rice, wheat and oat [10, 9, 3, 23, 22] could be achieved. However, the morphogenetic competence of cells is significantly reduced during long term maintenance and the phenomenon of somaclonal variation limits the suitability of these cells for transformation.

These limitations could be overcome by directly targeting tissues or cells which can be obtained easily and manipulated *in vitro*. In cereals, scutellar tissue of immature embryos, immature inflorescences or microspores are suitable primary explants for bombardment or tissue electroporation as it was reported in maize, wheat, barley, rice, tritordeum and triticale [16, 26, 24, 2, 19, 5, 13, 25, 1, 27]. The time necessary for preparation of the target cells is comparatively low and the risk of somaclonal variation is negligible as the period in tissue culture is reduced to a few weeks. Another advantage of microprojectile bombardment of primary explants is that even genotypes which are recalcitrant in protoplast culture can be transformed easily [5].

In the following pages we will present protocols for the isolation and transformation of scutellar tissue of immature embryos and the culture and selection procedure used to obtain fertile transgenic wheat plants.

Procedures

The primary requirement for an optimal target is that the tissue or cells receiving exogenous DNA are culturable *in vitro*, actively dividing and capable of giving rise to fertile plants. In our experiments we used scutellar tissue of immature embryos of the winter wheat genotype "Florida" and the spring type line "Veery" as a target for particle bombardment. The *in vitro* culture system used, allowed plant regeneration at high frequencies. Optimization of the transformation parameters was performed by transient transformation experiments using the plasmid pDB1 [2], containing the *gus* gene under control of the actin1-promoter from rice [17] and the selectable marker *bar* driven by the CaMV 35S-promoter. The aim of these experiments was to enhance transient transformation by minimizing tissue damage, which is correlated with a reduced regeneration capability.

For the selection of putative transgenic plants we have developed an *in vitro* selection system based on the resistant gene *bar* which confers resistance against the herbicide BASTA. The advantage of herbicide selection is not only the possibility to select plants *in vitro*, but also to identify transgenic plants at each time point of development by spraying plants with a herbicide solution.

Putative transgenic regenerants, transferred into soil, were analyzed for the presence of GUS-activity histochemically and for PAT-activity by spraying plants with a BASTA herbicide solution.

Isolation of immature embryos and preparation for bombardment

Steps in the procedure
1. Obtain spikes bearing immature caryopses 12–14 days after pollination.
2. Remove the immature caryopses and sterilize for 1 min in 70% ethanol and for 20 min in 1% NaOCl, 0.5 % Mucasol.
3. Wash three times with sterile distilled water.
4. Dissect the embryos in a sterile environment and place scutellum-side up on modified L3D2 [13] callus induction medium (without amino acids). For particle bombardment, place 20–30 embryos in the center of a 9 cm Petri dish.
5. Seal the culture dishes with Parafilm and incubate in the dark at 26 °C.
6. The isolated embryos can be bombarded directly after isolation or after a one to two days preculture.

Notes
1. The developmental stage of the immature embryos is the most significant factor which regulates the response *in vitro*. Best response (high capability of somatic embryogenesis and plant regeneration) is obtained from embryos in which the morphological development has been nearly completed, and the deposition of starch in the scutellar tissue has just begun.
6. A preculture treatment prior to bombardment depends on the genotype used. For example, in our experiments scutellar tissue of the winter wheat genotype "Florida" showed no significant differences in culture response without a preculture prior to bombardment, whereas in the case of immature embryos from the spring type line "Veery" only a preculture of one or, preferably, two days gave the same frequencies of somatic embryogenesis as non-bombarded controls. Using longer preculture treatments, we never obtained transgenic plants.

Preparation of DNA coated gold particles

Steps in the procedure
1. Place 40 mg gold particles (0.4–1.2 μm) per 1 ml of 96 % ethanol in a microtube. Sonicate the particle suspension for 30 sec using a standard tip to destroy particle aggregates. Centrifuge the microtube for 1 min at 4200 × *g* in a microfuge, remove the supernatant and add 1 ml 96 % ethanol. Resuspend the particles briefly and repeat three times.
2. Wash particles three times in 1 ml sterile distilled water as described previously. After the last centrifugation step, resuspend particles in 1 ml sterile distilled water. Aliquot 50 μl of the final suspension into microtubes, while vortexing the suspension continually. Store aliquots at −20 °C.
3. For precipitation of plasmid DNA, add to a 50 μl aliquot, 5 μl of plasmid DNA (1 μg/μl) and vortex briefly.
4. Add 50 μl of a 2.5 M solution of calcium chloride and 20 μl of a 0.1 M spermidine (free base) solution. The suspension should be vortexed while adding each solution. Incubate on ice for 15 min.
5. Spin the microcarriers down in a microfuge at 4200 × *g* for 5 sec. and remove the supernatant. Wash particles with 250 μl of absolute ethanol by vortexing for 1 min, centrifuge in the same manner and remove the supernatant.
6. Resuspend particles in 240 μl of absolute ethanol. Particles attached to the microtube wall can be scraped and resuspended thoroughly with a pipet.

Notes
1. Gold particles with an average size of 0.4–1.2 μm are available from Heraeus, Karlsruhe, Germany.
6. DNA coated microprojectiles should be used for bombardment immediately after preparation.

Bombardment of scutellar tissue

The particle gun employed in these experiments was a DuPont PDS 1000/He gun. The bombardment parameters used are summarized in Table 1.

Steps in the procedure
1. Clean the PDS 1000/He particle delivery system, the sample chamber and all the other material used for bombardment with 70 % ethanol. Allow time for drying in a sterile environment.
2. Seat a macrocarrier into the macrocarrier holder. Pipet 3.5 μl of the freshly prepared DNA-microcarrier suspension in the centre of the macrocarrier. The particle suspension must briefly be vortexed each time. Dry for 2 min.
3. Place a stopping screen in the stopping screen support and the macrocarrier holder containing the macrocarrier on the top rung of the fixed nest. Fix the macrocarrier holder with the macrocarrier cover lid. Place in position.
4. Place a rupture disk of the desired burst pressure in the recess of the rupture disk retaining cap and screw the rupture disk retaining cap onto the gas acceleration tube.
5. Put a Petri dish containing the immature embryos on the Petri dish holder and place in position.
6. Close the sample chamber door and evacuate to a partial vacuum of 27 inch Hg.
7. Press the FIRE switch to allow pressure to build in the acceleration tube.
8. After the rupture disk has burst release the vacuum in the sample chamber.

Table 1. Successfully used parameters for particle bombardment of scutellar tissue of wheat

Distance between	
– rupture disk and macrocarrier	2.5 cm
– macrocarrier and stopping screen	0.8 cm
– stopping screen and target cells	5.5 cm
gas pressure	900–1550 psi
partial vacuum	27 inch Hg
particles	gold, 0.4–1.2 μm
particle amount per bombardment	29 μg

An average number of 100 transient transformation events per embryo could be observed using a plasmid construct containing the *gus* marker gene driven by the actin-1 promoter [17]. The number of transient events was reduced, using greater distances between the stopping screen and the target cells and/or lower partial vacuum. No significant differences in transient transformation numbers were observed using helium gas pressures between 900–1550 psi. Lower or higher gas pressures gave reduced transient numbers and, in the case of higher gas pressures, higher tissue damage. The important step in minimizing tissue damage was the reduction of the amount of particles used per bombardment. We observed a direct correlation between the particle amount (between 29 and 116 μg per bombardment) used for bombardment and the capability of bombarded tissue to develop somatic embryos in high frequency. On the other hand, there were no significant differences in the number of transient transformation events. Using 29 μg particles per bombardment, the same rate of somatic embryogenesis and plant regeneration could be observed as in non-bombarded controls.

Transient expression assay

Following bombardment culture embryos for two days at 26 °C in the dark. For histochemical detection of GUS-activity [14, 15], overlay bombarded embryos with x-Gluc staining buffer. Incubate for 12–18 h at 37 °C. Count the number of blue spots.

Culture, selection and plant regeneration

Steps in the procedure

1. Spread the embryos over the culture plate, one day after bombardment. Developing coleoptiles from the embryo must be cut off in the following days.
2. Subculture developing calluses 2 weeks after transformation for additional 14 days on L3D2/B3 selection medium. At this time point, the development of embryoid-like structures are visible on the surface of the developing calluses.
3. Transfer callus showing the development of somatic embryos after 14 days on L3D0.1Z10/B3 regeneration medium for shoot development. Culture under light conditions (3000 lx for 16 h) at 26 °C.
4. Subculture after additional 14 days on L3D0.1/B3 regeneration medium for root development. Subculture at 14–21 d intervals on the same medium.
5. Transfer rooted plantlets with a leaf length of 1.5–2.0 cm to half strenght MS-medium for additional 14 days.
6. Transfer rooted plants to a peat/soil mix and grow under greenhouse conditions to maturity.

Notes

1.   The further development of the coleoptile can be observed in embryos of an older developmental stage. The induction of somatic embryogenesis is in principle possible, but it takes place at lower frequency. The development of somatic embryos on the scutellum is inhibited when the coleoptile develops.
2.   For the selection of transgenic plants, we use the herbicide BASTA. The solution contains 20% phosphinothricin, the active component of the herbicide.
3. and 4.   The same selection pressure used during the callus induction phase was also used during plant regeneration.
5.   Plantlets with healthy shoot systems which do not form roots *in vitro* may sometimes be rooted by placing a sterile droplet of a 1 mg/ml solution of IBA (Indole-3-butyric acid) to the leaf bases.
6.   The plantlets must be hardened in a high humidity chamber for about 1 week before transfer to the greenhouse.

Identification of transgenic regenerants

After the transfer of putative transgenic regenerants into soil, all plants are analyzed for enzyme activity of both introduced genes. GUS-activity is monitored histochemically and PAT-activity indirectly by spraying plants with a BASTA-solution which is toxic for non transformed plants.

Steps in the procedure

A. Detection of GUS-activity in leaf segments
1. Harvest leaf pieces of 1 cm in length and incubate them in GUS-assay buffer for 12 to 16 h at 37 °C.
2. Extract the chlorophyll by incubating leaf pieces in a 3 : 1 solution of ethanol : glacial acetic acid for 1 h at 70 °C.
3. GUS-activity is visible under the microscope.

B. Detection of PAT-activity
4. 14 days after transfer of plants into soil, spray whole plants or only single leaves with an aqueous solution of the herbicide BASTA (150 mg/l PPT, 0.1% Tween 20).
5. Examine plants one week after herbicide application for necrosis. Resistant plants do not show necrosis or only partial necrosis at the leaf tips, whereas sensitive plants do not survive herbicide treatment. As a negative control, use in each experiment non transformed regenerants of the same developmental stage.

Notes
3. The staining intensity depends on the expression level and the developmental stage of the leafs. Younger leaves normally show a stronger reaction than older ones.
5. The lethal dose of PPT depends on the developmental stage of the plants. Older plants tolerate higher concentrations than younger ones. In order to get clear results of the status of older plants it is necessary to use PPT concentrations between 200 and 250 mg/l.

Solutions

GUS assay buffer
– 5 mM potassium ferricyanide
– 5 mM potassium ferrocyanide
– 0.05% (w/v) 5-bromo-4-chloro-3-indolyl-β-D-glucuronic acid
– 0.06% (v/v) Triton X-100
– 0.2 M sodium phosphate buffer, pH 7.0

Sterilise by filtration. Store aliquots at -20 °C.

– Sigma 1-A agarose: prepare double-concentrated (1.6% for all media) in dis-
  tilled water. Autoclave befor using.
– PPT stock solution: 20 mg/l PPT (use the herbicide BASTA) in sterile dis-
  tilled water.
– DNA solution: dissolve DNA in sterile 10 : 1 mM Tris/HCl:EDTA buffer, pH
  8.0. Final concentration 1 μg/μl.
– Spermidine solution: 0.1 M in sterile distilled water. Store at -80 °C.
– CaCl$_2$ solution: 2.5 M in sterile distilled water. Store at -20 °C

References

1. Barcelo P, Hagel C, Becker D, Martin A, Lörz H (1994) Transgenic cereal (tritordeum) plants obtained at high efficiency by microprojectile bombardment of inflorescence tissue. The Plant Journal 5(4): 583–592.
2. Becker D, Brettschneider R, Lörz H (1994) Fertile transgenic wheat from microprojectile bombardment of scutellar tissue. The Plant Journal 5(2): 299–307.
3. Cao J, Duan X, McElroy D, Wu R (1992) Regeneration of herbicide resistant transgenic rice plants following microprojectile mediated transformation of suspension culture cells. Plant Cell Rep 11: 586–591.
4. Chan MT, Chang HH, Ho SL, Tong WF, Yu SM (1993) *Agrobacterium*-mediated production of transgenic rice plants expressing a chimeric alpha-amylase promoter/beta-glucuronidase gene. Plant Mol Biol 22: 491–506.
5. Christou P, Ford TL, Kofron M (1991) Production of transgenic rice (*Oryza sativa* L.) plants from agronomically important indica and japonica varieties via electrical discharge particle acceleration of exogenous DNA into immature zygotic embryos. Bio/Technol 9: 957–962.
6. D'Halluin K, Bonne E, Bossut M, De Beuckeleer M, Leemans J (1993) Transgenic maize plants by tissue electroporation. Plant Cell 4: 1495–1505.
7. Datta SK, Peterhans A, Datta K, Potrykus I (1990) Genetically engineered fertile indica-rice recovered from protoplasts. Bio/Technol 8: 736–740.
8. Donn G, Eckes P, Müller H (1992) Genübertragung auf Nutzpflanzen. BioEngineering 8: 40–46.
9. Fromm ME, Morrish F, Amstrong C, Williams R, Thomas J, Klein TM (1990) Inheritance and expression of chimeric genes in progeny of transgenic maize plants. Bio/Technol 8: 833–839.
10. Gordon-Kamm WJ, Spencer TM, Mangano ML, Adams TR, Daines RJ, Start WG, O'Brian JV, Chambers SA, Adams JWR, Willetts NG, Rice TB, Mackey CJ, Krueger W, Kausch AP, Lemaux PG (1990) Transformation of maize cells and regeneration of fertile transgenic plants. Plant Cell 2: 603–618.
11. Hiei Y, Ohta S, Komari T, Kumashiro T (1994) Efficient transformation of rice (Oryza sativa L.) mediated by Agrobacterium and sequence analysis of the boundaries of the T-DNA. The Plant Journal 6(2): 271–282.
12. Jähne A, Becker D, Brettschneider R, Lörz H (1994) Regeneration of transgenic microspore-derived, fertile barley. Theor Appl Genet 89: 525–533.
13. Jähne A, Lazzeri PA, Jäger-Gussen M, Lörz H (1991a) Plant regeneration from embryogenic cell suspensions derived from anther cultures of barley (*Hordeum vulgare* L.). Theor Appl Genet 82: 74–80.
14. Jefferson RA (1987a) Assaying chimeric genes in plants: The GUS gene fusion system. Plant Mol Biol Rep 5: 387–405.
15. Jefferson RA, Kavanagh TA, Bevan MW (1987b) GUS fusions: β-glucuronidase as a sensitive and versatile gene fusion marker in plants. EMBO J 6: 3901–3907.
16. Koziel GM, Beland GL, Bowman C, Carozzi NB, Crenshaw R, Crossland L, Dawson J, Desai N, Hill M, Kadwell S, Launis K, Lewis K, Maddox D, McPherson K, Meghji MR, Merlin E, Rhodes R, Warren GW, Wright M, Evola SV (1993) Field performance of elite transgenic maize plants expressing an insecticidal protein derived from *Bacillus thuringiensis*. Bio/Technol 11: 194–200.
17. McElroy D, Zhang W, Cao J, Wu R (1990) Isolation of an efficient actin promoter for use in rice transformation. Plant Cell 2: 163–171.
18. Murashige T, Skoog F (1962) A revised medium for rapid growth and bioassays with tobacco tissue cultures. Physiol Plant 15: 473–497.

19. Nehra NS, Chibbar RN, Leung N, Caswell K, Mallard C, Steinhauer L, Baga M, Kartha K (1994) Self-fertile transgenic wheat plants regenerated from isolated scutellar tissues following microprojectile bombardment with two distinct gene constructs. The Plant Journal 5(2): 285–297.
20. Sanford JC, Klein TM, Wolf ED, Allen N (1987) Delivery of substances into cells and tissues using a particle bombardment process. J Part Sci Technol 5: 27–37.
21. Shimamoto K, Terada R, Izawa T, Fujimoto H (1989) Fertile transgenic rice plants regenerated from transformed protoplasts. Nature 338: 2734–276.
22. Somers DA, Rines HW, Gu W, Kaeppler HF, Bushnell W R (1992) Fertile, transgenic oat plants. Bio/Technol 10: 1589–1594.
23. Vasil V, Castillo AM, Fromm ME, Vasil IK (1992) Herbicide resistant fertile transgenic wheat plants obtained by microprojectile bombardment of regenerable embryogenic callus. Bio/Technol 10: 667–674.
24. Vasil V, Srivastava V, Castillo AM, Fromm ME, Vasil IK (1993) Rapid production of transgenic wheat plants by direct bombardment of cultured immature embryos. Bio/Technol 11: 1553–1558.
25. Wan Y, Lemaux PG (1994) Generation of large numbers of independently transformed fertile barley plants. Plant Physiol 104: 37–48.
26. Weeks JT, Anderson OD, Blechl AE (1993). Rapid production of multiple independent lines of fertile transgenic wheat (*Triticum aestivum*). Plant Physiol 102: 1077–1084.
27. Zimny J, Becker D, Brettschneider R, Lörz H (1995). Fertile, transgenic Triticale (x *Triticosecale* Wittmack). Mol Breeding 1, No.2: 155–164.

Plant Tissue Culture Manual **B13**, 1–46, 1996.

Transient gene expression and stable genetic transformation into conifer tissues by microprojectile bombardment

ARMAND SÉGUIN, DENIS LACHANCE & PIERRE J. CHAREST

Molecular Genetics and Tissue Culture Group; Petawawa National Forestry Institute; Box 2000, Chalk River, Ontario K0J 1J0, Canada; E-mail: aseguinpnfi.forestry.ca

Introduction

Genetic tranformation technologies are essential to programs of molecular biology and genetic engineering. Although significant efforts in conifer molecular biology have been initiated since the late 1980s, so far only a limited number of genes have been cloned [1]. Conifers are by far the most difficult plant group for this type of study because of their large genomes and lengthy life cycles. Furthermore, progress has been hindered by the present inefficiencies in gene transfer methods and tissue culture protocols for certain species such as pines. Nevertheless, there is a significant body of literature on gene transfer in several conifer species using *Agrobacterium*-mediated transformation and other protocols of direct DNA transfer.

For example, specific strains of *Agrobacterium* that induce tumors on seedlings have been identified and, in some cases, the transfer and integration of T-DNA encoded genes has been confirmed (see Table 1 and ref. [2, 3, 4] for a complete list). The only success for transgenic tree regeneration was reported with *Larix decidua* [5] but the number of trees produced was limited to less than a dozen [6]. There is no report of *Agrobacterium* transformation using a conifer tissue culture system.

Table 1. *Agrobacterium*-mediated DNA transformation of conifers

Genus	Observations (results)	Reference
Abies	Tumours obtained on seedlings; opines detected	[7, 8]
Larix	Roots observed on in vitro *grown stem*	[9, 10, 11]
	Transgenic trees regenerated from inoculated seedlings	[5, 6]
Libocedrus	Tumours obtained on seedlings	[12]
Picea	Tumours obtained on seedlings; opines detected	[7, 13, 14, 15, 16]
Pinus	Tumours obtained on plants and seedlings; NPT II transgene expression, Southern blot analysis	[8, 11, 12, 13, 16, 17, 18, 19, 20]
Pseudotsuga	Tumours obtained on seedlings; opines detected; NPT II transgene expression; Southern blot analysis	[8, 12, 13, 21, 22]
Taxus	Tumours obtained on shoot segments; opines detected; Southern blot analysis	[23]
Tsuga	Tumours obtained on seedlings	[8]

NPT II = neomycin phosphotransferase.

With direct DNA transfer methods, transient gene expression of the delivered genes has been obtained with electroporation, polyethylene glycol DNA delivery, silicon carbide-mediated DNA delivery and microprojectile bombardment (Table 2). This last method has emerged as simple and promising for the stable genetic transformation of conifers and other recalcitrant species, and has been used to regenerate transgenic plants of black spruce (*Picea mariana*), white spruce (*Picea glauca*), and tamarack (*Larix laricina*) [24, 25]. Furthermore, microprojectile DNA delivery has been an invaluable tool for studying expression and regulation in conifers of various genes from both angiosperms and gymnosperms (Table 2 and ref. [2]). It provides a tool to bypass the long life cycle of conifers by allowing gene delivery and expression in mature tree tissues such as flowers, pollen, differentiating wood, and needles [26].

We shall describe the protocols employed in our laboratory for direct DNA transfer in black spruce using microprojectile-mediated DNA delivery and we shall also describe the gene expression assay procedures used (β-glucuronidase,

Table 2. Direct DNA transformation in conifers

Genus	Method used and observations	Reference
Larix	Transient expression of GUS and CAT genes by electroporation of protoplasts	[27]
	Transient expression of GUS gene by microprojectile bombardment of somatic embryogenic tissues	[28, 29]
	Stable transformation by microprojectile bombardment of somatic embryogenic tissues	[25]
Picea	Transient expression of GUS and CAT genes by electroporation or PEG-mediated expression of protoplasts	[30, 31, 32, 33]
	Transient expression of GUS gene by silicon carbide-mediated DNA delivery	[34]
	Transient expression of GUS gene by microprojectile bombardment of somatic embryogenic tissues	[28, 35, 36, 37, 38, 39, 40]
	Stable transformation by microprojectile bombardment of somatic embryogenic tissues	[24, 25, 41, 42]
Pinus	Transient expression of Luc or CAT genes by electroporation of protoplasts	[32, 43]
	Transient expression of GUS gene by microprojectile bombardment of cell suspension, cotyledons, and differentiating wood	[43, 44, 45, 46]
Pseudotsuga	Transient expression of GUS gene by microprojectile bombardment of cotyledons	[47]

CAT= chloramphenicol acetyltransferase, GUS=β-glucuronidase, Luc= firefly luciferase, PEG= polyethylene glycol.

neomycin phosphotransferase, and luciferase assays). The microprojectile DNA delivery method can be used with any plant tissue but this section will only cover protocols for somatic embryogenic tissues and for pollen.

With conifers, somatic embryogenesis is an ideal tissue culture system for gene transfer experiments because it can be induced readily from tree tissues (immature and mature zygotic embryos, cotyledons, and needles from young seedlings) and plants can be regenerated from tissue culture lines [48, 49]. Two stages of somatic embryogenesis were used for transient gene expression and stable genetic transformation; mature somatic embryos and embryonal masses. In transient gene expression studies, our laboratory has tested over 35 different chimeric gene constructs for level of expression and tissue specificity. Several factors affecting the level of expression of introduced reporter genes using particle bombardment in *Picea* embryogenic tissues have been described previously. For instance, the choice of tissue line and the time kept in culture will result in variation in transient gene expression of the reporter gene [35, 37, 38]. Moreover, the type of vector used and the strength of the promoter driving the reporter gene have also been shown to be important [35, 36, 38].

For stable gene transfer, we have obtained transgenic tissue culture lines using kanamycin selection for black spruce, white spruce, and tamarack. From these, transgenic trees have been regenerated in black spruce and tamarack and, depending on the effort invested, unlimited numbers of transgenic trees may be obtained. The procedure yields trangenic tissues at low frequency, but can be repeated consistently. Some of the transgenic tissue cultures have been maintained for 3 years without loss in the level of foreign gene expression. Our laboratory is improving transgenic embryogenic line recovery by investigating (a) the use of other selective agents (e.g., geneticin), (b) by designing vectors carrying scaffold or matrix attachment regions [50] and; (c) by altering the physiology of the tissue culture lines. We have already tested selection using hygromycin and methotrexate resistance genes as markers but with no success.

It is difficult to estimate the frequency of stable transformant recovery in relation to the level of transient gene expression obtained. According to several authors, a large variation can be observed in the estimation of the conversion rates from transiently expressing cells to stably transformed cells when using microprojectile-DNA delivery in plant cells as detected by GUS histochemical staining. The data can vary from a conversion rate of approximately 1% [51] to 5% of the cells that transiently expressed a foreign gene and then stably integrated it [52]. For the conifer species with which we regenerated transgenic lines, the ratio of cells showing transient gene expression to the number of lines stably transformed is lower than 0.1%.

For pollen, microprojectile bombardment has been used successfully to achieve transient gene expression in tobacco [53, 54], lily [54, 55], and maize [56] for the study of tissue-specific gene expression. Pollen grains have also been proposed as target material for gene transfer in plants [57, 58]. Because pollen grains from conifers are usually relatively large in diameter (about 50–100 μm for spruce), are easy to collect in large quantities, and can be preserved for an extended period of time, we have investigated the genetic transformation of pollen using microprojectile bombardment technology [40, 59]. Comparison of the germination frequency between bombarded pollen expressing GUS activity and non-bombarded pollen showed that pollen vigour does not seem affected by the bombardment procedure [40]. The frequency of gene transfer into this tissue without optimization as indicated by transient gene expression is in the order of 5–8% [59]. These data are encouraging from the perspective of using microprojectile bombardment of pollen for the stable transformation of conifer germlines.

Protocols for gene delivery in pollen and embryogenic tissues of black spruce

The protocols that follow were used to optimize gene delivery, as indicated by transient gene expression, and to assess the strength of different gene constructs driven by various promoters. Although optimization of gene transfer for transient gene expression is often considered a first step in determining the parameters needed for stable transformation, conditions for microprojectile bombardment used for stable transformation differed from those for transient gene expression. We describe here protocols for black spruce and we give some indications of the conditions that can differ for white spruce and tamarack. Furthermore, these protocols can be applied to other spruce and larch species using different tissue culture media.

Procedures

i– Tissue culture protocols

The tissue culture protocols for black spruce embryogenic cultures including initiation of embryogenic cultures, maintenance, maturation of somatic embryos, germination, and transfer to soil have been described in Lelu *et al.* [49] and in Cheliak and Klimaszewska [60]. Furthermore, chapter C3 (pp. 1–16) of this manual, by Thorpe and Harry [61], covers somatic embryogenesis of conifers. The embryogenic tissue culture lines of black spruce used in this procedure were induced from mature seeds.

Embryonal masses
Embryogenic cell suspension cultures are established by transferring approximately 5 g of embryonal masses from gelled 1/2 LM medium to 40 ml liquid 1/2 LM medium. The suspensions are maintained by weekly transfer of embryogenic suspension into fresh medium (ratio 1 : 1) for a total volume of 40 ml in 250 ml Erlenmeyer flasks. These flasks are kept on a gyratory shaker at 120 rpm under indirect light with a 16h photoperiod. Cell suspension cultures of embryonal masses of black spruce are preferable for transformation although embryonal masses maintained on solid media can be broken up in liquid media just prior to bombardment (see note). Actively dividing four-day-old suspensions are used for transformation experiments. In some cases, pretreatment of embryogenic tissues before bombardment in liquid medium with increasing osmoticum can result in enhanced expression of the reporter gene [40]. Augmentation of the osmoticum medium may induce plasmolysis of the cells, making them less likely to release protoplasm after the penetration of the microprojectiles into the cells. Similar results, using particle bombardment, have been obtained with white spruce embryogenic material and suspension-cultured cells of tobacco [62] and maize [63] using mannitol, sorbitol, or raffinose as the osmoticum.

Somatic embryos
Somatic embryos are produced by placing embryonal masses on 1/2LM maturation media. Mature cotyledonary somatic embryos (size: 2–3 mm) are produced after 4–8 weeks at which stage they are used for targets for microprojectile bombardment.

Notes
1. Higher levels of transient gene expression are observed with cell suspension cultures because they are more easily and uniformly spread than callus on solid media.

ii– Microprojectile bombardment and reporter gene assays

A– Microprojectile preparation

Steps in the procedure

1. In a 1.5 ml microfuge tube, place 60 mg of gold microprojectiles.
2. Add 1 ml absolute ethanol and vortex for 1–2 minutes. Allow tube to stand for 1 hr with brief periodic vortexing.
3. Briefly centrifuge the tube and remove supernatant. Wash the microprojectiles two times with 1 ml sterile distilled water by repeating resuspension and centrifugation. Finally, resuspend in sterile distilled water to a final volume of 500 μl.
4. Aliquot 25 μl of gold microprojectile suspension into 0.5 ml microfuge tubes.
5. Add 10 μg of vector DNA at a concentration of 1 μg/μl in water or TE and mix thoroughly.
6. While vortexing the DNA/gold microprojectile mixture, add 50 μl of 2.5 M CaCl$_2$ and 20 μl of 0.1M spermidine (free base). Continue vortexing for 30 sec. then let stand for 10 minutes.
7. Briefly centrifuge the tube and remove supernatant.
8. Add 200 μl of absolute ethanol and vortex. Briefly spin down microprojectiles and remove ethanol.
9. Resuspend the microprojectiles in 50 μl of absolute ethanol. 5 μl of this suspension (1 μg of DNA) will be used for each bombardment.
10. Use the DNA/ microprojectile preparations as soon as possible.

Notes

1. Gold microprojectiles with a diameter of 1.6 μm are from Bio-Rad Laboratories, Richmond, CA.
3. The gold microprojectiles may be stored at room temperature for up to a month.
5. Plasmid DNA was isolated by alkaline lysis [64] and subsequently purified on Qiagen™ anion exchange resin according to the protocol provided by the manufacturer (Qiagen, Chatsworth, CA) or by CsCl gradient [65]. The optimal concentration of plasmid DNA to be added to the gold microprojectiles for maximum transient transformation can be established by a dose-response curve. However, increasing gene delivery with higher DNA quantities will reach a plateau and eventually lead to a decrease in transformation efficiency, presumably due to inappropriate conditions for DNA precipitation on the gold microprojectiles. Aggregation of the gold microprojectiles is often observed at high DNA concentrations and this reduces efficiency of cell penetration and could cause cell injuries.
6. Plasmid DNA is adsorbed onto the gold microprojectiles using the procedure described originally by Klein [52].

B– Tissue bombardment

DNA transfer is carried out using the Biolistic™ Particle Delivery system PDS-1000/He System (referred to as the "gene gun" in the following pages; Bio-Rad Laboratories, Richmond, CA) following the manufacturer's recommendations and as described in Kikkert [66] (See note). The various settings and parameters used are given in Table 3.

Table 3. Potential settings of the PDS-1000/He apparatus and parameters used for transient expression

PDS-1000/He parts	Tissue bombarded			
	Settings	Pollen	Embryonal masses	Somatic embryos
Rupture Disc pressure (PSI)	400	O		
	650			
	900		X	
	1100	×	×	×
	1300			
	1550			
	1800		O	
	2000			
	2200			
Gap distance (cm)	0.32			
	0.64	×		
	0.95	O	×, O	×
Fixed nest assembly (macrocarrier flight distance, mm)	3	×		
	8	O	O	
	13		×	×
Sample holder distance (cm)	6			×
	9	×, O	×, O	
	12			

×=settings used for black spruce
O=settings used for white spruce

Steps in the procedure
1. Sterilize stop screens, kapton discs, and disc holders in ethanol (70%) and then allow to air dry.
2. Position the kapton disc within the disc holder.
3. Pipet 5 μl of the DNA preparation from the tube (while vortexing) and deposit in the center of the kapton disc. Let dry under a laminar flow until ethanol has evaporated.
4. Position a rupture disc in the rupture disc holder and screw tightly into position.

5. Assemble the fixed nest as per the manufacturers' instructions and position in the gene gun housing.
6. Place the target tissue on the sample holder beneath the nest assembly and close the gene gun door.
7. Evacuate chamber with a vacuum of 675 mm Hg (26.5 inch Hg) and fire the gene gun.
8. Release vacuum slowly and remove the bombarded sample.
9. Repeat steps 2–8 for each sample.

Notes
1. Our laboratory compared the PDS-1000 gunpowder system with the helium system. Also, the tungsten microprojectiles were tested in both systems. The helium system and the gold microprojectiles consistently gave better results. The stop screens, kapton discs, and disc holders should be resterilized before each use.

C– Reporter gene assays

Standard protocols to detect expression of the β-glucuronidase gene [67], neomycin phosphotransferase II gene, and luciferase gene [68] are suitable for enzymatic assays in conifer tissues. Small modifications have been made to prevent interference by endogenous biochemicals such as the addition of methanol for the fluorescent GUS assay and increase in the concentration of phosphate buffer in the luciferase assay for black spruce.

β-glucuronidase assay:

Histochemical staining

Steps in the procedure
1. Completely submerge tissue to be assayed in GUS histochemical buffer.
2. Incubate material in the dark at 37 °C for 24 hours.
3. Count the number of cell clusters (GUS expression units) in embryonal masses or of pollen grains with a blue colouration, indicating GUS gene expression.

Note
1. The GUS histochemical buffer can be vacuum infiltrated for 5 min for increased sensitivity.

Fluorometric assay

Steps in the procedure
1. Harvest tissue and place in 1.5 ml microfuge tube.
2. Grind or sonicate tissue in 200 μl of GUS extraction buffer (GEB).
3. Pellet the cell debris by centrifugation.
4. Take 100 μl sample from the clarified extract and combine with 100 μl GEB containing 2 mM MUG (4-methyl umbelliferyl β-D-glucuronide) and 40% methanol in a new microfuge tube.
5. After incubation at 37 °C for 1, 2, and 4 hours, take out aliquots of 20 μl and stop the reaction by adding to the aliquot to 1980 μl of 0.2 M Na_2CO_3.
6. Determine the fluorescence for each sample and calculate GUS activity as described in Jefferson [69] by comparing with a standard curve using MU (4-methyl umbelliferone) as a standard.

Notes
4. As described in Kosugi [70], we found that addition of methanol in the reaction buffer is essential to remove endogenous fluorescence of spruce and larch, tissues probably due to high phenolic content.
6. In our laboratory, the fluorescence was determined in a TKO-100 fluorometer (Hoeffer Scientific Instrument, San Francisco, CA). A dose-response curve can be obtained with both the histochemical staining and the fluorogenic assays. In several publications using microprojectile bombardment technology, the quantitative determination of GUS gene activity is done by assessing the number of discrete areas of blue histochemical staining. Determination of the transient GUS activity in pmole of 4-methylumbelliferyl β-D-glucuronide (MUG) per minute per mg of protein extracted facilitate the evaluation of the efficiency of the method used.

Neomycin phosphotransferase assay

The neomycin phosphotransferase gene product is assayed using an ELISA (enzyme linked immunosorbent assay) kit from 5 Prime → 3 Prime, Inc. (Boulder, CO, USA). The steps are as indicated by the manufacturer and the extraction buffer used is the one described for plant tissues (0.25M Tris-Cl, pH 7.8, 1.0 mM phenylmethylsulfonylfluoride). Alternatively, the fluorometric GUS extraction buffer can be substituted when both GUS and NPT II are being used to quantify relative gene expression of two different promoters.

Firefly luciferase assay

This assay is done according to the luciferase assay kit from Promega Biotech (Fisher Scientific, Ottawa, Ontario, Canada). Modifications to the extraction buffer and procedure were required due to luciferase enzyme inhibition with black spruce tissues.

Steps in the procedure
1. Collect the bombarded tissues 36 hours after bombardment and add 100 μl of cell culture lysis buffer.
2. Freeze the mixture in liquid nitrogen and grind to produce an uniform powder or, in an Eppendorf tube, grind directly the sample in cell culture lysis buffer.
3. Centrifuge cells and add 20 μl of supernatant to 100 μl of beetle luciferin (470 μM) in reaction buffer.
4. Count immediately the number of photons emitted by using a scintillation counter (specially equipped) or a luminometer. Express the data as 10^6 photon events/mg of protein.

Protein assay

Protein concentration of the extracts is evaluated with the Bio-Rad Bradford Protein assay kit or the Bio-Rad *DC* protein assay kit (Bio-Rad, Mississauga, Ont., Canada) following manufacturer's protocols.

iii– Gene transfer for transient gene expression into conifer pollen

A– Pollen collection and storage

Steps in the procedure
1. At the first sign of dehiscence, collect and bag sections of twigs containing several microstrobili (male flower cones) directly from the field.
2. Release the pollen by shaking bags and collect by pouring into scintillation vials.
3. Store black spruce and white spruce pollen desiccated at $-20\ °C$.

Notes
1. The male flower can also be forced to dehiscence under controlled laboratory conditions.
3. The pollen can be stored for up to a year with marginal loss in germination rate.

B– Pollen bombardment

Steps in the procedure

1. Weigh out the desired amount of pollen.
2. Make a suspension of pollen in sterile distilled water. 5 ml of the suspension should contain the desired amount pollen per bombardment.
3. Vacuum filter 5 ml of suspension onto a nylon membrane using a sintered glass filter assembly to ensure uniform distribution.
4. Place the nylon filter in the center of a 9 cm Petri dish containing solidified pollen germination media or three layers of Whatman filter papers soaked with 1.5 ml of the same liquid media.
5. Prepare DNA-coated gold microprojectiles for transformation (see section ii A).
6. Bombard pollen within 30 minutes of plating.
7. Incubate the pollen at 24 °C until it germinates.
8. Assay the pollen for expression of the transferred gene(s).

Notes

3. Nylon membrane (e.g., for nucleic acid hybridization) is preferred as the pollen can be easily washed or scraped off and recovered if desired. Do not use a Buchner funnel as the pollen tends to be unevenly distributed on the filter. Vigorous stirring of the suspension is required to maintain homogeneity. For transient assays, various amounts of pollen (1 to 5 mg) are used.
7. 16–24 hours is sufficient for black spruce.

iv– Gene transfer for transient gene expression in embryogenic tissues

A– Embryonal masses

Steps in the procedure
1. Vacuum filter 0.1–0.5 g black spruce cell suspension culture onto paper filter discs (5.5 cm diam, Whatman # 2) using a Buchner funnel.
2. Position the filter in the center of a 9 cm diameter Petri dish containing solidified 1/2LM maintenance media.
3. Prepare vector DNA for transformation.
4. Bombard the target tissue using the PDS1000/He.
5. Incubate the tissue at 24 °C for 48hrs in the dark.
6. Assay tissue for expression of the introduced genes.

B– Somatic embryos

Steps in the procedure

1. Position a 3 cm × 3 cm piece of nylon mesh (200 μm pore size) in the center of a 9 cm diameter Petri dish containing solidified 1/2LM maintenance media. Carefully position (disperse horizontally) 30 to 35 mature somatic embryos on the nylon mesh.
2. Prepare vector DNA for transformation.
3. Bombard the target somatic embryos using PDS1000/He.
4. Incubate the bombarded somatic embryos at 24 °C in the dark for 48 hours.
5. Assay the somatic embryos for the expression of the introduced genes.

Notes

1. The embryos are placed on a fine mesh to prevent them from becoming embedded in the media and also for ease of handling.

v– Gene transfer for stable transformation of embryogenic tissues

A– Determination of the antibiotic concentration for optimal selection.

The antibiotic concentration to be used for selection of tranformed tissues is determined by establishing toxicity curves and by identifying a concentration that will cause inhibition of growth but not complete killing. This was essential for spruce and larch embryogenic cultures selected for kanamycin resistance.

Embryonal masses

Steps in the procedure
1. Dilute a 4-day-old cell suspension culture in 1/2 LM maintenance media to a concentration of 20 mg/ml.
2. Vacuum filter 5 ml of diluted suspension (100 mg of embryonal masses) onto paper filter disc (5.5 cm diam. Whatman #2) using a Buchner funnel.
3. Transfer the filter bearing the cells on a Petri dish containing solidified 1/2 LM maintenance media.
4. Weigh the filter paper and the cells after one week of incubation and transfer them on fresh 1/2 LM media containing a given concentration of kanamycin (8 Petri dishes with each of the following concentrations: 0, 10, 15, 20, 25, and 30 μg/ml).
5. At weekly intervals, the filter paper discs from each Petri dish are weighed and subcultured to fresh media each 2 weeks.
6. The lowest concentration that inhibits growth and results in cell mortality after 4 weeks on selection should be chosen (viability is assessed by vital staining or growth).

Somatic embryos

Steps in the procedure
1. Place somatic embryos (16) in a Petri dish containing 1/2 LM gelled media with a given concentration of kanamycin (0, 5, 7.5, 10, 12.5 and 15 μg/ml).
2. Incubate in the dark at 24 °C for 8 weeks with a subculture after 4 weeks.
3. Identify minimal (threshold) antibiotic concentration that inhibits secondary embryogenic growth but does not kill all regrowth.

B– Stable transformation using embryonal masses as target material.

Steps in the procedure
1. Tissue preparation and bombardment procedures are the same as those used for transient gene expression.
2. After bombardment, incubate tissues on a filter paper placed on gelled 1/2 LM medium in the dark for 7 days at 24 °C.
3. Then transfer the bombarded tissues on the filter paper to maintenance media containing the selective agent (25 μg/ml kanamycin for black spruce and tamarack). Return to previous incubation conditions for 6–8 weeks. Subculture to fresh media after 4 weeks.
4. Identify putatively transformed embryonal masses by screening any growing tissue for expression (enzymatic assay) or presence of the introduced gene by polymerase chain reaction (PCR) [71]. After 8 weeks, remove all tissues from the selective media and place on regular maintenance media and continue to monitor for any growth.
5. Bulk up transformed embryonal masses then pass through liquid media containing 500–750 μg/l of kanamycin for two weeks to eliminate or reduce chimerism.
6. Maintain transgenic lines on selection free media or, preferably, cryopreserve the tissues [72, this manual] to avoid loss of regeneration capacity or change in gene expression.
7. Place tissues on maturation media to produce somatic embryos that will germinate to regenerate transgenic trees. For this, follow the established procedure described in Lelu [49] and Thorpe and Harry [61, this manual].

C– Stable transformation using mature somatic embryos as target material.

Steps in the procedure
1. Tissue preparation and tissue bombardment procedures are the same as those used in transient expression.
2. Incubate the bombarded somatic embryos on a filter paper deposited on gelled media at 24 °C in the dark for 7days.
3. After this time, spread embryos out over the surface of fresh maintenance media containing the selective agent (10 μg/ml Km- no mesh required) and return to the previous incubation conditions for 6–8 weeks.
4. Identify putatively transformed callus formed through secondary somatic embryogenesis by screening all secondary embryogenic growth for expression (enzymatic assay) or presence of the introduced genes (PCR).
5. Bulk up the transformed callus and then briefly pass (2 weeks) through liquid media containing 500–750μg/l kanamycin to eliminate or reduce chimerism.
6. Maintain the transgenic lines on selection free media or preferrably cryopreserve the transgenic lines.
7. Tissues can be matured to produce somatic embryos from which transformed plantlets can be obtained.

Note
4. Secondary somatic embryogenesis is defined as the re-induction of embryonal masses from somatic embryos using the same conditions as for the induction of embryonal masses from zygotic embryos. The tissue culture medium used for induction of somatic embryogenesis is the same as used for maintenance.

Solutions:

For 1 litre of half strength Litvay's medium (1/2LM; [41])

1/2 LM 5X frozen stock	200 ml
Casein hydrolysate (casamino acids)	1 g
Sucrose (1% final)	10 g
2,4-D (1mg/ml stock solution)	2.2 ml
6-benzylaminopurine (BA, 0.5mg/ml stock solution)	2.2 ml
Add distilled H_2O to 1 1	
pH	5.7

Autoclave and add 20 ml of filter sterilized glutamine (25 mg/ml stock solution) to cooled medium.

If solid medium is required, add gelrite (4g/l) before adjusting pH and before autoclaving.

Maturation medium (1 litre)
same as 1/2LM but replace 2,4-D and BA with abscisic acid

Abscisic acid (10 mM stock solution)	2.0 ml
Sucrose (6% final)	60 g

For 2 litres of Litvay's 5X stock (1/2 LM 5X frozen stock)

NH_4NO_3	8.21 g
KNO_3	9.5 g
$MgSO_4.7H_2O$	9.25 g
KH_2PO_4 (monobasic)	1.7 g
$CaCl_2.2H_2O$	0.11 g
LM micro nutrient stock (100X)	100 ml
LM vitamin stock (100X)	100 ml
Myo-inositol	1 g
Fe diethylene triamine pentaacetate	0.4 g

Add distilled H_2O to 2000 ml and store frozen in 100 ml aliquots for further use.

For 1 litre of LM micronutrient stock 100X

KI	0.415 g
H_3BO_3	3.1 g
$MnSO_4.H_2O$	2.1 g
$ZnSO_4.7H_2O$	4.3 g
$Na_2MoO_4.2H_2O$	0.125 g
$CuSO_4.5H_2O$	0.05 g
$CoCl_2.6H_2O$	0.013 g

Add distilled H_2O to 1000 ml and store frozen in 100 ml aliquots for further use.

For 1 litre of lm vitamin stock 100X

Nicotinic acid	0.05 g
Pyridoxine HCl	0.01 g
Thiamine HCl	0.01 g

Add distilled H_2O to 1000ml and store frozen in 100 ml aliquots for further use.

GUS histochemical buffer
0.5 mg/ml X-GLUC (5-bromo-4-chloro-3-indolyl glucuronide)
100 mM sodium phosphate pH 7.0
0.5 mM ferrocyanide
0.5 mM ferricyanide
0.5% (V/V) Triton X-100
1 mM EDTA

GUS extraction buffer (GEB)
50 mM NaHPO4 pH 7.0
10 mM β-mercaptoethanol
0.5 M Na_2EDTA pH 8.0
10% Triton X-100

Luciferase assay cell lysis buffer
100 mM Tris-phosphate pH 7.8
2 mM DTT
2 mM DACT
10% glycerol
1 % Triton

Luciferase assay reaction buffer
20 mM Tricine
1.07 mM $((MgC)_3)_4MG(OH)_2$
267 mM $MgSO_4$
0.1 mM EDTA
33 mM DTT
270 μM Coenzyme A
530 μM ATP pH7.8

Pollen germination medium (Brewbaker [73])
8.0 mM H_3BO_3
1.3 mM $Ca(NO_3)_2$
1.0 mM KNO_3
1.5 mM $MgSO_4$
5% (W/V) sucrose
pH of 5.2
0.6% agar

References

1. Charest PJ, Rutledge R (1993) Is genetic engineering a viable option for tree improvement? In: Proceedings of the twenty-fourth meeting of the Canadian Tree Improvement Association, pp 27–40, CTIA proceedings.
2. Ellis DD (1992) Transformation in *Picea*. In: Bajaj YPS (Ed.) Biotechnology in Agriculture and Forestry, Springer-Verlag, Berlin and New York (in press).
3. Manders G, Davey MR, Power JB (1992) New Genes for Old Trees. J Exp Bot 43: 1181–1190.
4. van Wordragen MF, Dons HJM (1992) *Agrobacterium tumefaciens*-mediated transformation of recalcitrant crops. Plant Mol Biol Rep 10: 12–36.
5. Huang Y, Diner AM, Karnosky DF (1991) *Agrobacterium rhizogenes*-mediated genetic transformation and regeneration of a conifer: *Larix decidua*. In Vitro Cell Dev Biol 27P: 201–207.
6. Karnosky DF, Podila GK, Tsai CJ, Chiang VL, Shin D-I (1994). Progress in production of transgenic trees with value-added genes: Results with larch and aspen. In: Biological Sciences Symposium, pp 157–160, Tappi Press, Minneapolis, Minnesota.
7. Clapham DH, Ekberg I (1986) Induction of tumours by various strains of *Agrobacterium tumefaciens* on *Abies nordmanniana* and *Picea abies*. Scand J For Res 1: 435–437.
8. Morris JW, Castle LA, Morris RO (1989) Efficacy of different *Agrobacterium tumefaciens* strains in transformation of pinaceous gymnosperms. Physiol Mol Plant Pathol 34: 451–461.
9. Diner AM, Karnosky DF (1987) Differential responses of two conifers to *in vitro* inoculation with *Agrobacterium rhizogenes*. Eur J For Path 17: 211–216.
10. Karnosky DF, Diner AM, Barnes WM (1988) A model system for gene transfer in conifers: european larch and *Agrobacterium*. In: Somatic cell genetics of woody plants, Ahuja MR (Ed.), pp 55–63, Kluwer Academic Publishers, Dordrecht, The Netherlands.
11. McAfee BJ, White EE, Pelcher LE, Lapp MS (1993) Root induction in pine (*Pinus*) and larch (*Larix*) spp. using *Agrobacterium rhizogenes*. Plant Cell Tiss Org Cult 34: 53–62.
12. Stomp A-M, Loopstra CA, Chilton WS, Sederoff RR, Moore LW (1990) Extended host range of *Agrobacterium tumefaciens* in the genus *Pinus*. Plant Physiol 92: 1226–1232.
13. Ellis DD, Roberts D, Sutton B, Lazaroff W, Webb D, Flinn B (1989) Transformation of white spruce and other conifer species by *Agrobacterium tumefaciens*. Plant Cell Rep 8: 16–20.
14. Clapham D, Ekberg I, Eriksson G, Hood EE, Norell L (1990) Within-population variation in susceptibility to *Agrobacterium tumefaciens* A281 in *Picea abies* (L.) Karst. Theor Appl Genet 79: 654–656.
15. Hood EE, Clapham DH, Ekberg I, Johannson T (1990) T-DNA presence and opine production in turmors of *Picea abies* (L.) Karst induced by *Agrobacterium tumafaciens* A281. Plant Mol Biol 14: 111–117.
16. Magnussen D, Clapham D, Grönroos R, von Arnold S (1994) Induction of hairy and normal roots on *Picea abies*, *Pinus sylvestris* and *Pinus cortorta* by *Agrobacterium rhizogenes*. Scand J For Res 9: 46–51.
17. Sederoff R, Stomp A-M, Chilton WS, Moore LW (1986) Gene transfer into loblolly pine by *Agrobacterium tumefaciens*. Bio/Technol 4: 647–650.
18. Gupta PK, Dandekar AM, Durzan DJ (1988) Somatic proembryo formation and transient expression of a luciferase gene in Douglas-fir and loblolly pine protoplasts. Plant Sci 58: 85–92.
19. Loopstra CA, Stomp A-M, Sederoff RR (1990) *Agrobacterium*-mediated DNA transfer in sugar pine. Plant Mol Biol 15: 1–9.
20. Bergmann BA, Stomp A-M (1992) Effect of host plant genotype and growth rate on *Agrobacterium tumefaciens*-mediated gall formation in *Pinus radiata*. Phytopathol 82: 1457–1462.
21. Dandekar AM, Gupta PK, Durzan DJ, Knauf V (1987) Transformation and foreign gene expression in micropropagated Douglas-fir (*Pseudotsuga menziesii*). Bio/Technol 5: 587–590.

22. Morris JW, Morris RO (1990) Identification of an *Agrobacterium tumefaciens* virulence gene inducer from the pinaceous gymnosperm *Pseudotsuga menziesii*. Proc Nat Acad Sci USA 87: 3614–3618.
23. Han KH, Fleming P, Walker K, Loper M, Chilton WS, Mocek U, Gordon MP, Floss HG (1994) Genetic transformation of mature *Taxus* – an approach to genetically control the in vitro production of the anticancer drug, taxol. Plant Sci 95: 187–196.
24. Ellis DD, McCabe DE, McInnis S, Ramachandran R, Russell DR, Wallace KM, Martinell BJ, Roberts DR, Raffa KF, McCown BH (1993) Stable transformation of *Picea glauca* by particle acceleration. Bio/Technol 11: 84–89.
25. Charest PJ, Lachance D, Devantier Y, Klimaszewska KK (1994) Transient gene expression and stable genetic transformation in *Picea mariana* (black spruce) and *Larix laricina* (tamarack). Rev Invest Agraria: Serie Syst Rec For 4: 187–192.
26. Charest PJ, Calero N, Lachance D, Mitsumune M, Yoo BY (1993) The use of microprojetile DNA delivery to bypass the long life cycle of tree species in gene expression studies. In: Current topics in Botanical Research vol. 1, Menon J (Ed.), Council of Scientific Research Integration, pp 151–163, India.
27. Charest PJ, Devantier Y, Ward C, Schaffer U, Klimaszewska KK (1991) Transient expression of foreign chimeric genes in the gymnosperm hybrid larch following electroporation. Can J Bot 69: 1731–1736.
28. Duchesne LC and Charest PJ (1992) Effect of promoter sequence on transient expression of the β-glucuronidase gene in embryogenic calli of *Larix* x *eurolepsis* and *Picea mariana* following microprojection. Can J Bot 70: 175–180.
29. Duchesne LC, Lelu MA, von Aderkas P, Charest PJ (1993) Microprojectile-mediated DNA delivery in haploid and diploid embryogenic cells of *Larix* spp. Can J For Res 23: 312–316.
30. Bekkaoui F, Pilon M, Laine E, Raju DSS, Crosby WL, Dunstan DI (1988) Transient gene expression in electroporated *Picea glauca* protoplasts. Plant Cell Rep 7: 481–484.
31. Bekkaoui F, Datla RSS, Pilon M, Tautorus TE, Crosby WL, Dunstan DI (1990) The effects of promoter on transient expression in conifer cell lines. Theor Appl Genet 79: 353–359.
32. Tautorus TE, Bekkaoui F, Pilon M, Datla RSS, Crosby WL, Fowke LC, Dunstan DI (1989) Factors affecting transient gene expression in electroporated black spruce (*Picea mariana*) and jack pine (*Pinus banksiana*) protoplasts. Theor Appl Genet 78: 531–536.
33. Wilson SM, Thorpe TA, Moloney MM (1989) PEG-mediated expression of GUS and CAT genes in protoplasts from embryogenic suspension cultures of *Picea glauca*. Plant Cell Rep 7: 704–707.
34. Charest PJ, Lachance D, Jones C, Devantier Y (1993) Microprojectile and silicon carbide mediated DNA delivery in conifers and recovery of transgenic black spruce. In Vitro Cell Devel Biol 29A: 87.
35. Duchesne LC, Charest PJ (1991) Transient expression of the β-Glucuronidase gene in embryogenic callus of *Picea mariana* following microprojection. Plant Cell Rep 10: 191–194.
36. Ellis DD, McCabe D, Russel D, Martinell B, McCown BH (1991) Expression of inducible angiosperm promoters in a gymnosperm, *Picea glauca* (white spruce). Plant Mol Biol 17: 19–27.
37. Newton RJ, Yibrah HS, Dong N, Clapham DH, von Arnold S (1992) Expression of an abscisic acid responsive promoter in *Picea abies* (L.) Karst. following bombardment from an electric discharge particle accelerator. Plant Cell Rep 11: 188–191.
38. Charest PJ, Calero N, Lachance D, Datla RSS, Duchesne LC, Tsang EWT (1993) Microprojectile-DNA delivery in conifer species – factors affecting assessment of transient gene expression using the β-Glucuronidase reporter gene. Plant Cell Rep 12: 189–193.
39. Bommineni VR, Datla RSS, Tsang EWT (1994) Expression of *gus* in somatic embryo cultures of black spruce after microprojectile bombardment. J Exp Bot 45: 491–495.
40. Li Y-h, Tremblay FM, Séguin A (1994) Transient transformation of pollen and embryogenic tissues of white spruce (*Picea glauca* (Moench.) Voss) resulting from microprojectile bombardment. Plant Cell Rep 13: 661–665.

41. Robertson D, Weissinger AK, Ackley R, Glover S, Sederoff RR (1992) Genetic transformation of norway spruce (*Picea abies* (L) Karst) using somatic embryo explants by microprojectile bombardment. Plant Mol Biol 19: 925–935.

42. Bommineni VR, Chibbar RN, Datla RSS, Tsang EWT (1993) Transformation of white spruce (*Picea glauca*) somatic embryos by microprojectile bombardment. Plant Cell Rep 13: 17–23.

43. Campbell MA, Kinlaw CS, Neale DB (1992) Expression of luciferase and β-glucuronidase in *Pinus radiata* suspension cells using electroporation and particle bombardment. Can J For Res 22: 2014–2018.

44. Stomp A-M, Weissinger A, Sederoff RR (1991) Transient expression from microprojectile-mediated DNA transfer in *Pinus taeda*. Plant Cell Rep 10: 187–190.

45. Loopstra CA, Weissinger AK, Sederoff RR (1992) Transient gene expression in differentiating pine wood using microprojectile bombardment. Can J For Res 22: 993–996.

46. Walter C, Smith DR, Connett MB, Grace L, White DWR (1994) A biolistic approach for the transfer and expression of a *gus*A reporter gene in embryogenic cultures of *Pinus radiata*. Plant Cell Rep 14: 69–74.

47. Goldfarb B, Strauss SH, Howe GT, Zaerr JB (1991) Transient gene expression of microprojectile-introduced DNA in Douglas-fir cotyledons. Plant Cell Rep 10: 517–521.

48. Tautorus TE, Fowke LC, Dunstan DI (1991) Somatic embryogenesis in conifers. Can J Bot 69: 1873–1899.

49. Lelu MA, Klimaszewska KK, Jones C, Ward C, von Aderkas P, Charest PJ (1993). A laboratory guide to somatic embryogenesis in spruce and larch. No. PI-X-111. Petawawa National Forestry Institute.

50. Allen GC, Hall GE, Childs LC, Weissinger AK, Spiker S, Thompson WF (1993) Scaffold attachment regions increase reporter gene expression in stably transformed plant cells. Plant Cell 5: 603–613.

51. Gordon-Kamm WJ, Spencer TM, Mangano ML, Adams TR, Daines RJ, Start WG, O'Brien JV, Chambers SA, Adams WRJ, Willetts NG, Rice TB, Mackey CJ, Krueger RW, Kausch AP, Lemaux PG (1990) Transformation of maize cells and regeneration of fertile transgenic plants. Plant Cell 2: 603–618.

52. Klein TM, Harper EC, Svab Z, Sanford JC, Fromm ME, Maliga P (1988) Stable genetic transformation of intact *Nicotiana* cells by the particle bombardment process. Proc Natl Acad Sci USA 85: 8502–8505.

53. Twell D, Klein TM, Fromm ME, McCormick S (1989) Transient expression of chimeric genes delivered into pollen by microprojectile bombardment. Plant Physiol 91: 1270–1274.

54. Nishihara M, Ito M, Tanaka I, Kyo M, Ono K, Irifune K, Morikawa H (1993) Expression of the β-glucuronidase gene in pollen of lily (*Lilium longiflorum*), tobacco (*Nicotiana tabacum*), *Nicotiana rustica*, and peony (*Paeonia lactiflora*) by particle bombardment. Plant Physiol 102: 357–361.

55. van der Leede-Plegt LM, van de Ven BCE, Bino RJ, van der Salm TPM, van Tunen AJ (1992) Introduction and differential use of various promoters in pollen grains of *Nicotiana glutinosa* and *Lilium longiflorum*. Plant Cell Rep 11: 20–24.

56. Hamilton D, Roy M, Rueda J, Sindhu R, Sanford J, Mascarenhas J (1992) Dissection of a pollen-specific promoter from maize by transient transformation assays. Plant Mol Biol 18: 211–218.

57. Negrutiu I, Heberle-Bors E, Potrykus I (1986) Attempts to transform for kanamycin-resistance in mature pollen of tobacco. In: Mulcahy DL, Bergamini-Mulcahy G, Ottaviano E (Eds.) Biotechnology and Ecology of Pollen, pp 65–70, Springer-Verlag, New York.

58. Twell D, Klein TM, McCormick S (1991) Transformation of pollen by particle bombardment. In: Lindsey K (Ed.) Plant Tissue Culture Manual, pp 1–14, Kluwer Academic, Dordrecht.

59. Hay I, Lachance D, von Aderkas P, Charest PJ (1994) Transient chimeric gene expression in pollen of five conifer species following microparticle bombardment. Can J For Res 24: 2417–2423.

60. Cheliak WM, Klimaszewska KK (1991) Genetic variation in somatic embryogenesis response in open-pollinated families of black spruce. Theor Appl Genet 82: 185–190.
61. Thorpe TA, Harry IS (1991) Clonal propagation of conifers. In: Lindsey K (Ed.) Plant Tissue Culture Manual, C3: pp 1–16, Kluwer Academic Publishers, Dordrecht.
62. Russell JA, Roy MK, Sanford JC (1992) Major improvements in biolistic transformation of suspension-cultured tobacco cells. In Vitro Cell Dev Biol-Plant 28P: 97–105.
63. Vain P, McMullen MD, Finer JJ (1993) Osmotic treatment enhances particle bombardment-mediated transient and stable transformation of maize. Plant Cell Rep 12: 84–88.
64. Birnboim HC, Doly J (1979) A rapid alkaline extraction procedure for screening recombinant plasmid DNA. Nucleic Acids Res 7: 1513–1517.
65. Sambrook J, Fritsch EF, Maniatis T (1989) Molecular Cloning: A Laboratory Manual (2nd ed.), Cold Spring Harbor, NY: Cold Spring Habor Laboratory.
66. Kikkert JR (1993) The Biolistic PDS-1000/He device. Plant Cell Tissue Organ Cult 33: 221–226.
67. Jefferson RA (1987) Assaying chimeric genes in plants: The GUS gene fusion system. Plant Mol Biol Rep 5: 387–405.
68. Howell SH, Ow DW, Schneider M (1989) Use of the firefly luciferase gene as a reporter of gene expression in plants. In: Gelvin SB, Schilperoort RA (Eds.) Plant Molecular Biology Manual, pp 1–11, Kluwer Academic Publishers, Dordrecht.
69. Jefferson RA, Kavanagh TA, Bevan MW (1987) GUS fusion: β-Glucuronidase as a sensitive and versatile gene fusion marker in higher plants. EMBO J 6: 3901–3907.
70. Kosugi S, Ohashi Y, Nakajima K, Arai Y (1990) An improved assay for β-glucuronidase in transformed cells: methanol almost completely suppresses a putative endogenous β-glucuronidase activity. Plant Science 70: 133–140.
71. Saiki RK, Gelfand DH, Stoffel S, Scharf SJ, Higuchi R, Horn GT, Mullis KB, Erlich HA (1988) Primer-directed enzymatic amplification of DNA with a thermostable DNA polymerase. Science 239: 487–491.
72. Charest PJ, Bonga J, Klimaszewska K (1996) Cryopreservation of plant tissue cultures: the example of embryogenic tissues from conifers. In: Plant Tissue Culture Manual, Lindsey K (Ed.), C9: pp 1–27 Kluwer Academic Publishers, Dordrecht.
73. Brewbaker JL, Kwack BH (1963) The essential role of calcium ion in pollen germination and pollen tube growth. Am J Bot 50: 859–865.

Plant Tissue Culture Manual **C9**, 1–27, 1996.
© *1996 Kluwer Academic Publishers.*

Cryopreservation of plant tissue cultures: the example of embryogenic tissue cultures from conifers

P.J. CHAREST[1]*, J. BONGA[2] & K. KLIMASZEWSKA[1]

[1]*Molecular Genetics and Tissue Culture Group, Petawawa National Forestry Institute, Canadian Forest Service, NRCan, Chalk River, Ontario, Canada K0J 1J0, E-mail: pcharest@pfni.forestry.ca;* [2]*Maritimes Region Canadian Forest Service, NRCan, Fredericton, New Brunswick, Canada E3B 5P7 (*Author for correspondence)*

Introduction

Several methods are available to maintain plant tissue cultures. Each method has advantages and disadvantages regarding the labour required, the frequency of necessary transfers, the availability of equipment, the potential for contamination, the risk of somaclonal variation, and the type of tissue to be preserved [1]. Subculturing is used in most laboratories because of its simplicity. However, it is risky because of the possibility of contamination and somaclonal variation and it is not practical for long-term preservation because it is labour intensive. Typically, subculture is required every 1 to 4 weeks. Another method is reduction of the growth rate (often called slow growth or minimal growth), which is the same as subculturing except that growth of tissue is inhibited by special environmental conditions [2, 3]. Artificial seed production is also used to maintain plant tissue cultures. It can be used with mature somatic embryos desiccated to a level that arrests metabolism [4, 5]. The limitation of the method is that it can only be used in species where somatic embryogenesis is available and desiccation can be applied successfully. Both slow growth maintenance and artificial seed production are medium-term approaches (2 months to a year). The best method for long term storage is the freezing of tissues at the temperature of liquid nitrogen (-196 °C) or its gaseous phase (-140 °C) which suspends all metabolic processes. This technique, called cryopreservation, has four types of protocols [6]. In all of the protocols, the critical point is to avoid intracellular ice crystal formation by ensuring that the cells and tissues are adequately dehydrated. The first type of protocol is referred to as conventional slow freezing. It includes exposure to chemical cryoprotectants, followed by ice inoculation and gradual slow freezing to -40 °C. The tissues are then placed in liquid nitrogen. The second type is called simple freezing and includes exposure to cryoprotectants at room temperature, followed by rapid freezing at -30 °C (temperature of a domestic freezer). The vials containing the tissues are then placed in liquid nitrogen. The third type is vitrification where the tissues are treated with high concentrations of cryoprotectants and then directly placed in liquid nitrogen. The resulting solution (cellular content and cryoprotectants) is supercooled, which transforms it into amorphous glass without ice crystallization. The fourth protocol includes the desiccation of the plant tissues to induce physiological drought followed by

Table 1. Types of plant tissues that have been cryopreserved

Protocols	Type of plant tissues
Conventional slow freezing	cultured cells [10, 11, 12], meristems [13] and somatic embryos [14]
Simple freezing	cultured cells [15, 16] and somatic embryos [17, 18]
Vitrification	winter hardy meristems and buds [19], meristems [20], orthodox pollen [21], somatic embryos [22], cultured cells [23]
Desiccation followed by vitrification	somatic embryos [24], lateral buds [25], apical meristems [26], shoot tips [27], excised zygotic embryos from recalcitrant seeds [28], recalcitrant pollen [29]

Orthodox = desiccation tolerant, Recalcitrant = desiccation sensitive, references are only examples of the protocols that were applied.

immersion in liquid nitrogen. Table 1 summarizes the four types of methods used and their potential application to different types of tissues. For more details on the theory and practice of cryopreservation applied to plant cells the reader should consult Kartha [7], Dresser et al. [8], and Withers [9].

The present chapter will describe the conventional slow freezing method that has been applied extensively in our laboratories to somatic embryogenic cultures of conifers [30, 31, 32]. As indicated in Table 2, it has been used with several conifer species and cell lines (genotypes) from which trees were regenerated after thawing. Cryopreservation of conifer embryogenic tissue does not appear to be genotype-dependent [32]. Conifer embryogenic cultures have been preserved for a period of 3 years at PNFI and they are still viable after thawing.

Table 2. Cryopreservation of conifer embryogenic lines at the Canadian Forest Service

Species	No. of lines in cryopreservation	Location
Larix decidua	9	PNFI
L. decidua (haploid)	4	PNFI
Larix laricina	12	PNFI
Larix laricina (transgenic)	2	PNFI
Larix leptolepis	3	PNFI
Larix leptolepis (haploid)	1	PNFI
Larix x eurolepis	11	PNFI
Larix x leptoeuropaea	26	PNFI
L. occidentalis	2	PNFI
Picea abies	5	PNFI
Picea glauca	54	PNFI
	450	CFS- Maritimes Region
Picea glauca (transgenic)	19	PNFI
P. glauca engelmannii complex	16	PNFI
Picea mariana	259	PNFI
P. mariana (transgenic)	11	PNFI
Picea rubens	15	PNFI
Pinus strobus	11	PNFI
	3	CFS– Quebec Region
Total number of species and lines	12 species and over 913 lines	

No. of lines cryopreserved successfully as of March 1995

Procedures

Cryopreservation of conifer somatic embryogenic tissues comprises five main steps: choice of tissues to be frozen, chemical pretreatment, freezing, thawing, and regrowth [30]. The choice of tissue is critical for the success of the process. The protocol described below is optimized for tissues that are typically embryogenic (Fig. 1). As a general rule, rapidly growing tissues with a population of small cells that are

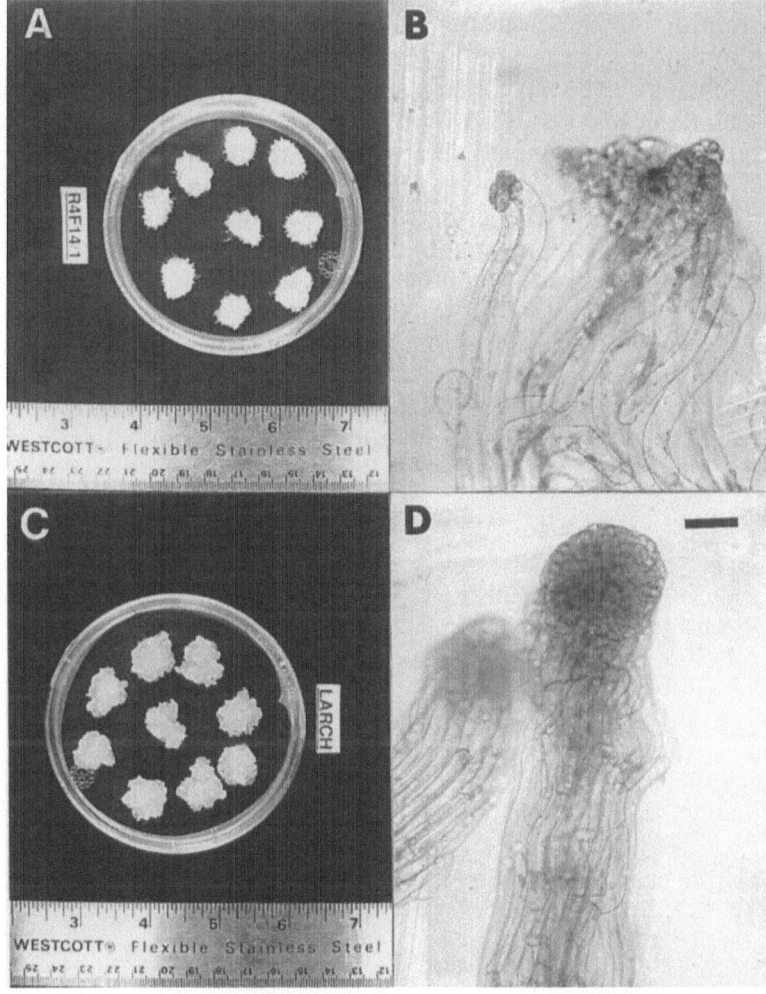

Fig. 1. Photograph of embryonal masses of A) *Picea mariana* and C) *Larix x eurolepis* at 2 weeks after subculture on maintenance media. Micrograph of embryonal masses showing isolated early embryos of B) *Picea mariana* and D) *Larix x eurolepis*. Two zones can be distinguished; the meristematic head (small dense cells) and the suspensor area (loose elongated cells). For B) and D), bar=32μm.

densely cytoplasmic will give the best results. One way to achieve rapid cell division is to use a cell suspension prepared from the embryonal masses.

Chemical pretreatment of the cells is required to dehydrate the tissues and to lower their freezing temperature. This is done by growing the tissues in media with high chemical osmoticum (e.g., polyethylene glycol or sorbitol) followed by a short treatment with a cryoprotectant such as dimethyl sulfoxide (DMSO). Following this treatment, the cells are gradually frozen from 0 °C to −40 °C at an average rate of −0.33 °C/min. This slow freezing is required to avoid internal ice crystal formation. Once the tissues have reached −40 °C, they are transferred to liquid nitrogen (either the gaseous phase at −160 °C or the liquid phase at −196 °C). Once frozen, there is no specific requirement with respect to storage. Regrowth of the frozen cells is accomplished by rapidly thawing the cells in a warm water bath (37 °C). The procedure is followed by a regrowth phase that includes a step to eliminate the osmoticum and the cryoprotectant. This reduces the osmotic shock and the toxicity associated with the cryoprotectant.

Typically, with spruce and larch species, only the meristematic cells (small densely cytoplasmic cells) survive the procedures. The large vacuolated cells of the embryo suspensor do not survive the process and a new suspensor is regenerated [31].

The protocol described here, or similar protocols, have been used in several other laboratories with conifer embryogenic tissue cultures [33–38].

Steps in the procedure

A) *Tissue culture protocols*

The tissue culture protocols for spruce and larch embryogenic cultures, including initiation of embryogenic cultures, maintenance, maturation of somatic embryos, germination, and transfer to soil have been described in Lelu et al. [39]. Furthermore, chapter C3 (pp. 1–16) of this manual by Thorpe and Harry [40] covers somatic embryogenesis of conifers. Table 2 gives a listing of species for which this protocol has been applied successfully. As examples, the recipes for *Picea mariana* and *Larix x eurolepis* for tissue culture maintenance and cryopreservation will be provided.

The embryonal masses of both species are maintained on solid medium (1/2 LM for *P. mariana* and MSG for *L. x eurolepis*) in Petri dishes (100 × 25mm; nine clumps per dish) and subcultured every 14 days. They can also be maintained in liquid medium (same media as above but without the gelling agent) in 125ml flasks containing 50ml of medium and subcultured every 7 days on a rotary shaker at 120 rpm.

B) *Material to be frozen*

The embryogenic tissue cultures should be in a vigorous phase of growth. Typically, 3-day-old embryonal masses from a cell suspension subcultured every 7 days or 4-day-old embryonal masses from a culture on solid medium subcultured every 7 or 14 days can be used.

Note
The embryogenic tissues should be monitored closely for their macroscopic and microscopic appearance. The manipulator must develop a knowledge of what constitutes a healthy tissue culture. Any changes could indicate that the line is not suitable for cryopreservation.

C) Pretreatment

1. If the cultures are maintained on gelled medium, suspend the embryonal masses in liquid tissue culture medium (1/2 LM for *P. mariana* and MSG for *L. x eurolepis*). Gently mix the suspended tissues and use a spatula if needed to break up large tissue pieces.
2. Collect the embryonal masses from the suspension by filtering through a nylon mesh (pore size 73 μm).
3. Transfer 1–2 g (fresh weight) of cells to 125 ml Erlenmeyer flasks each containing 7 ml of liquid medium (1/2 LM for *P. mariana* and MSG for *L. x eurolepis*) supplemented with 0.4 M sorbitol.
4. Place the flasks on a rotary shaker at 100–115 rpm for 20–24 hrs.
5. Place the flasks on ice.
6. Add stepwise, over a period of 30 min, 3 ml of a DMSO solution prepared in 0.4 M sorbitol medium so that the final concentration will reach 5–10% DMSO (volume/volume).
7. Mix gently and leave on ice for 30 min for the mixture temperature to equilibrate.

D) Freezing

8. Mix the pretreated suspension gently. Dispense 1.0 ml aliquots of suspended cells into 1.2 ml sterile cryogenic vials (Corning Lab. Science Co., USA).
9. Place vials in a programmable freezer. The freezer should be programmed to cool as follows: hold at 0 °C for 10 min and cool to -40 °C at a rate of 0.33 °C/min.
10. Remove the vials from the freezer and immerse in liquid nitrogen (-196 °C).

E) Thawing

11. Remove vials from storage and place them in a container with liquid nitrogen.
12. Prepare a warm water bath at 37–40 °C.
13. Remove vials from the liquid nitrogen, a few at a time, and thaw by swirling them in the warm water bath. The vials should be removed from the bath as soon as the ice pellet inside the vial has thawed, which should be within 2 min.
14. Allow the contents to equilibrate to room temperature (a few minutes) before opening the vial.

F) Regrowth

15. Pour contents of the vial onto two layers of sterile filter paper discs (Whatman #2, Whatman International Ltd., England) placed on the surface of the gelled medium in a Petri dish (1/2LM for *P. mariana* and MSG for *L. x eurolepis*).
16. Seal the Petri dish with parafilm and place in the dark at 25 °C.
17. After 24 hrs, transfer the upper filter paper disc with the cells to fresh medium to remove DMSO and excess liquid.
18. After 2 weeks, subculture as usual by transferring portions of the actively growing embryonal cell masses to a Petri dish containing gelled tissue culture medium (1/2 LM *for P. mariana* and MSG for *L. x eurolepis*).
19. Regenerate plantlets following the published protocols [39].

G) Vital staining of regrowing tissue cultures

The cell viability after thawing can be determined using fluorescein diacetate (FDA; [44]) by incubating the cells in 0.005% (weight/volume) of FDA for 30 minutes, followed by two washings in tissue culture medium. The cells are observed under a microscope with ultraviolet illumination. Living cells will produce fluorescence.

Notes
3. It is important that the embryonal cell mass density is such that after thawing there will be enough surviving cells with sustained growth. We found that increasing the fresh weight from 1 g to 2 g per 10 ml of the final medium volume will ensure a rapid regrowth in some lines.
10. Storage of the cryovials is preferable in the vapour phase of liquid nitrogen because this reduces the risks of the vials exploding. Ashwood-Smith and Friedmann [43] estimated that it would take up to 30 000 years before background radiation would have a significant effect on the genetic make up of the cultures stored at −196 °C.

Solutions

For 500 ml of half strength Litvay's medium (1/2LM; [41])

1/2 LM 5X frozen stock	100 ml
Casein hydrolysate (casamino acids)	0.5 g
Sucrose (1% final)	5.0 g
2,4-D (1 mg/ml stock solution)	1.1 ml
6-benzylaminopurine (0.5 mg/ml stock solution)	1.1 ml
Add distilled H_2O to 500 ml	
pH	5.7

Autoclave and add 10 ml of filter sterilized glutamine (25 mg/ml stock solution) to cooled medium.

If solid medium is required, add gelrite (2 g/500 ml) before adjusting pH and before autoclaving.

For 2 litres of Litvay's 5X stock (1/2 LM 5X frozen stock)

NH_4NO_3	8.25 g
KNO_3	9.5 g
$MgSO_4.7H_2O$	9.25 g
KH_2PO_4 (monobasic)	1.7 g
$CaCl_2.2H_2O$	0.11 g
LM micro nutrient stock (100X)	100 ml
LM vitamin stock (100X)	100 ml
Myo-inositol	1g
Fe diethylene triamine pentaacetate	0.28 g

Add distilled H_2O to 2000 ml and freeze in 100 ml sterile Twirlbags (Fisher Scientific Co, Canada) for further use.

For 1 litre of LM micronutrient stock 100X

KI	0.415 g
H_3BO_3	3.1 g
$MnSO_4.H_2O$	2.1 g
$ZnSO_4.7H_2O$	4.3 g
$Na_2MoO_4.2H_2O$	0.125 g
$CuSO_4.5H_2O$	0.05 g
$CoCl_2.6H_2O$	0.013 g

Add distilled H_2O to 1000 ml and freeze in 100 ml sterile Twirlbags for further use.

For 1 litre of LM vitamin stock 100X

Nicotinic acid	0.05 g
Pyridoxine HCl	0.01 g
Thiamine HCl	0.01 g

Add distilled H_2O to 1000 ml and freeze in 100 ml sterile Twirlbags for further use.

For 500 ml of MSG medium (MSG; [42])

MSG 5X frozen stock	100 ml
Sucrose (2%)	10.0 g
2,4-D (1mg/ml stock solution)	1.0 ml
6-benzylaminopurine	0.5 ml
Add distilled H_2O to 500 ml	5.8

Autoclave and add 29.2 ml of filter sterilized glutamine (25 mg/ml stock solution) to cooled medium.

If solid medium is required, add gelrite (2.0 g/500 ml) before autoclaving.

For 2 litres of MSG 5X stock

KNO_3	1.0 g
$CaCl_2.2H_2O$	4.4 g
$MgSO_4.7H_2O$	3.7 g
KH_2PO_4 (monobasic)	1.7 g
KCl	7.4 g
Fe diethylene triamine pentaacetate	0.4 g
MS micro nutrient stock (100X)	100 ml
MS vitamin stock (100X)	100 ml
KI (10 mg/ml stock solution)	0.83 ml

Add distilled H_2O to 2000 ml and freeze in 100 ml sterile Twirlbags for further use.

For 1 litre of MS micronutrient stock 100X

H_3BO_3	0.62 g
$MnSO_4.4H_2O$	2.23 g
ZnSO4.7H2O	0.86 g
$Na_2MoO_4.2H_2O$	0.025 g
$CuSO_4.5H_2O$	0.0025 g
$CoCl_2.6H_2O$	0.0025 g

Add distilled H_2O to 1000 ml and freeze in 100 ml sterile Twirlbags for further use.

For 1 litre of MS vitamin stock 100X

Myo-inositol	10.0 g
Nicotinic acid	0.05 g
Pyridoxine HCl	0.05 g
Thiamine HCl	0.01 g

Add distilled water to 1000 ml and freeze in 100 ml sterile Twirlbags for further use.

Concentrated dimethysulfoxide (DMSO) solution
For a solution of 10% DMSO in 10 ml, add 1.0 ml of tissue culture medium (1/2 LM for *P. mariana* and MSG for *L. x eurolepis*) with 0.4 M sorbitol to 1.0 ml double-strength tissue culture medium with 0.8 M sorbitol. Place on ice.

Gradually add 1.0 ml of filter sterilized DMSO and mix thoroughly. Make this solution just before adding to the pretreated cultures contained in 7.0 ml of tissue culture medium with 0.4 M sorbitol.

References

1. Wang BSP, Charest PJ, Downie B (1993) *Ex situ* storage of seeds, pollen and *in vitro* cultures of perennial woody plant species. FAO Forestry Paper #113. 83pp.
2. Banerjee N, de Langhe E (1985) A tissue culture technique for rapid clonal propagation and storage under minimal growth conditions of *Musa* (Banana and plantain). Plant Cell Rep. 4: 351–354.
3. Engelmann F (1990) Utilisation d'atmosphères à teneur en oxygène réduite pour la conservation de cultures d'embryons somatiques de palmier à huile (*Elaeis guineensis* Jacq.). C.R. Acad. Sci. Paris Série III 310: 679–684.
4. McKersie BD, Senaratna T, Bowley R, Brown DCW, Krochko JE, Bewley JD (1989) Application of artificial seed technology in the production of hybrid *alfalfa* (*Medicago sativa* L.). In Vitro Cell. Dev. Biol. 25: 1183–1188.
5. Redenbaugh K, Fujii JA, Slade D (1988) Encapsulated plant embryos. In: Biotechnology in Agriculture. A. Mizrahi ed. A.R. Liss New York, USA. pp. 225–248.
6. Sakai A (1993) Cryogenic strategies for survival of plant cultured cells and meristems cooled to −196 °C. JICA Group Ref. No. 6. pp. 5–26.
7. Kartha KK (1985) Cryopreservation of plant cells and organs. CRC Press Inc., Boca Raton, Florida. 276pp.
8. Dresser BL, Russell P, Pope CE, Pence V, Plair B, Long P (1992) Wildlife germplasm cryopreservation workshop manual. Center for Reproduction of Endangered Wildlife, Cincinnati Zoo & Botanical Garden, OH.
9. Withers LA (1980) Tissue culture storage for genetic conservation. International Board for Plant Genetic Resources Rep. Rome. 91pp.
10. Panis BJ, Withers LA, De Langhe EAL (1990) Cryopreservation of *Musa* suspension cultures and subsequent regeneration of plants. Cryo-Lett. 11: 337–350.
11. Gnanapragasam S, Vasil IK (1990) Plant regeneration from a cryopreserved embryogenic cell suspension of a commercial sugarcane hybrid (*Saccharum* sp.). Plant Cell Rep. 9: 419–423.
12. Wang ZY, Legris G, Nagel J, Potrykus I, Spangenber G (1994) Cryopreservation of embryogenic cell suspensions in *Festuca* and *Lolium* species. Plant Sci. 103: 93–106.
13. Reed BM (1990) Survival of in vitro-grown apical meristems of *Pyrus* following cryopreservation. Hort. Sci. 25: 111–113.
14. Bertrand-Desbrunais A, Fabre J, Engelmann F, Dereuddre J, Charrier A (1988) Reprise de l'embryogénèse adventive à partir d'embryons somatiques de caféier (*Coffea arabica* L.) après leur congélation dans l'azote liquide. C.R. Acad. Sci. Paris. Série III 307: 795–801.
15. Nishizawa S, Sakai A, Amano Y, Matsuzawa T (1992) Cryopreservation of asparagus (*Asparagus officinalis* L.) embryogenic suspension cells and subsequent plant regeneration by a simple freezing method. Cryo-Lett. 13: 379–388.
16. Sakai A, Kabayashi S, Oiyama I (1991) Cryopreservation of nucellar cells of navel orange (*Citrus sinensis* Osb.) by a simple freezing method. Plant Sci. 74: 243–248.
17. Tessereau H, Lecouteux C, Florin B, Schlienger C, Petiard V. (1991) Use of a simplified freezing process and dehydration for the storage of embryogenic cell lines and somatic embryos. Rev. Cytol. Biol. Végé. Bot. 14: 297–310.
18. Lecouteux C, Florin B, Tessereau H, Bollon H, Petiard V (1991) Cryopreservation of carrot somatic embryos using a simplified freezing process. Cryo-Lett. 12: 319–328.
19. Sakai A (1984) Cryopreservation of apical meristems. Hort. Rev. 6: 357–372.
20. Towill LE (1990) Cryopreservation of isolated mint shoot tips by vitrification. Plant Cell Rep. 9: 178–180.
21. Jorgensen J (1990) Conservation of valuable gene resources by cryopreservation in some forest tree species. J. Plant Physiol. 136: 373–376.
22. Engelmann F, Dereuddre J (1988) Cryopreservation of oil palm somatic embryos: importance of the freezing process. Cryo-Lett. 9: 220–235.

23. Sakai A, Kabayashi S, Oiyama I (1990) Cryopreservation of nucellar cells of navel orange (*Citrus sinensis Osb. var. brasiliensis* Tanada) by vitrification. Plant Cell Rep. 9: 30–33.
24. Dereuddre J, Tannoury M, Hassen N, Kaminski M, Vintejoux C (1992) Application de la technique d'enrobage à la congélation d'embryons somatiques de carotte (*Daucus carota* L.) dans l'azote liquide: étude cytologique. Bull. Soc. Bot. Fr. Lettres Bot. 139: 15–33.
25. Uragami A, Sakai A, Nagai M (1990) Cryopreservation of dried axillary buds from plantlets of *Asparagus officinalis* L. grown *in vitro*. Plant Cell Rep. 9: 328–331.
26. Poissonnier M, Monod V, Pagues M, Dereuddre J (1992) Cryopreservation dans l'azote liquide d'apex d'*Eucalyptus gunnii* (Hook. F.) cultivé *in vitro* après enrobage et déshydratation. Ann. Rech. Sylvicoles. Afocel. pp. 6–23.
27. Paulet F, Engelmann F, Glaszmann JC (1993) Cryopreservation of apices of *in vitro* plantlets of sugarcane (*Saccharum* sp. hybrids) using encapsulation/dehydration. Plant Cell Rep. 12: 525–529.
28. Pence VC (1990) Cryostorage of embryo axes of several large-seeded temperate tree species. Cryobiol. 27: 212–218.
29. Marchant R, Power JB, Davey JM, Chartier-Hollis JM, Lynch PT (1993) Cryopreservation of pollen from two rose cultivars. Euphytica 66: 235–241.
30. Charest PJ, Klimaszewska, K (1995) Cryopreservation of germplasm of *Larix* and *Picea* Species. In: Y.P.S. Bajaj (Ed.) Biotechnology in Agriculture and Forestry, vol. 32 Cryopreservation of plant germplasm – I, pp. 191–203. Springer-Verlag Berlin Heidelberg.
31. Klimaszewska K, Ward C, Cheliak M (1992) Cryopreservation and plant regeneration from embryogenic cultures of larch (*Larix x eurolepis*) and black spruce (*Picea mariana*). J. Exp. Bot. 43: 73–79.
32. Park YS, Pond SE, Bonga JM (1994) Somatic embryogenesis in white spruce (*Picea glauca*): genetic control in somatic embryos exposed to storage, maturation treatments, germination, and cryopreservation. Theor. Appl. Genet. 89: 742–750.
33. Gupta PK, Durzan DJ, Finkle BJ (1987) Somatic polyembryogenesis in embryogenic cell masses of *Picea abies* (Norway spruce) and *Pinus taeda* (loblolly pine) after thawing from liquid nitrogen. Can. J. For. Res. 17: 1130–1134
34. Laine E, Bade P, David A (1992) Recovery of plants from cryopreserved embryogenic cell suspensions of *Pinus caribaea*. Plant Cell Rep. 11: 295–298.
35. Kartha KK, Fowke LC, Leung NL, Caswell KL, Hakman I (1988) Induction of somatic embryos and plantlets from cryopreserved cell cultures of white spruce (*Picea glauca*). J. Plant Physiol. 132: 529–539.
36. Toivonen PMA, Kartha KK (1989) Cryopreservation of cotyledons of nongerminated white spruce [*Picea glauca* (Moench) Voss] embryos and subsequent plant regeneration. J. Plant Physiol. 134: 766–768.
37. Bercetche J, Galerne M, Dereuddre J (1990) Augmentation des capacités de régénération de cals embryogènes de *Picea abies* (L.) Karst après congélation dans l'azote liquide. C.R. Acad. Sci. Paris. Série III 310: 357–363.
38. Galerne M, Dereuddre J (1987) Survie de cals embryogènes d'épicea après congélation à −196 °C. Ann. Rech. Sylvicoles Afocel. pp. 7–32.
39. Lelu MA, Klimaszewska KK, Jones C, Ward C, von Aderkas P, Charest PJ (1993) A laboratory guide to somatic embryogenesis in spruce and larch. Petawawa National Forestry Institute Information Report PI-X-111. 57pp.
40. Thorpe TA, Harry IS (1991) Clonal propagation of conifers. Plant Tiss. Cult. Man. C3: 1–16.
41. Litvay JD, Johnson MA, Verma D, Einspahr D, Weyrauch K (1981) Conifer suspension culture medium development using analytical data from developing seeds. Inst. Pap. Chem. Tech. Pap. Serv. 115: 1–17.
42. Becwar MR, Nagmani R, Wann SR (1990) Initiation of embryogenic cultures and somatic embryo development in loblolly pine (*Pinus taeda*). Can. J. For. Res. 20: 810–817.

43. Ashwood-Smith MJ, Friedman JB (1979) Lethal and chromosomal effects of freezing, thawing, storage time and X-irradiation on mammalian cells preserved at -196 °C in dimethylsulfoxide. Cryobiol. 16: 132–140.
44. Wildholm SM (1972) The use of fluorescein diacetate and phenosafranine for determining viability of cultured plant cell. Stain Technol. 47: 189–193.

Plant Tissue Culture Manual **F2**, 1–43, 1996.
© 1996 *Kluwer Academic Publishers.*

In vitro culture, mutant selection, genetic analysis and transformation of *Physcomitrella patens*

DAVID COVE

Department of Genetics, University of Leeds, Leeds LS26 8JF, UK

Introduction

The suitability of mosses for the study of plant development and genetics has been appreciated for a considerable time [19]. Early studies extended to a number of species, including *Funaria hygrometrica* and *Physcomitrella patens*. More recent studies, using both classical and molecular genetic analysis to study development, have concentrated principally on *Physcomitrella* and most of the techniques described in this chapter have been developed for this species. However, some of these methods are suitable, with little modification, to other moss species, in particular to *Ceratadon purpureus*.

The main phase of the life cycle of mosses is the haploid gametophyte. Spore germination or tissue regeneration gives rise to a system of cell filaments, the protonema, from which are produced gametophores, the leafy shoots which comprise the more familiar part of most mosses. Gametes are produced on the gametophores. *Physcomitrella* is monoecious and both male and female gametes are produced on the same gametophore. Gamete fusion leads to the production of the diploid sporophyte which in turn produces haploid spores following meiosis. Conventional genetic analysis is possible but many developmentally-abnormal mutant strains are infertile and so must be analysed by parasexual methods, using somatic hybrids obtained by protoplast fusion.

The protonemal phase of the *Physcomitrella* life cycle, comprises two types of cell, chloronema and caulonema. Filaments of both cell types extend by the serial division of the apical cell. Sub-apical cells of both chloronemal and caulonemal filaments can also divide, usually no more than twice, to produce further filaments. Chloronemal filaments develop following spore germination or as a result of regeneration from all types of tissue. The apical cells of chloronemal filaments, which have a cell cycle of about 24 h under standard conditions (see below), may after a few days of growth, give rise to the second cell type, caulonema. The apical cells of caulonemal filaments, under comparable conditions, have a cell cycle time of only about six hours, and so comprise the adventitious phase of protonemata. Branching of the sub-apical cells of caulonemal filaments can give rise to three types of side branches. Most commonly, chloronemal filaments are produced. Less commonly, side branches may develop into either further caulonemal filaments, or gametophores.

PIPS 99928

In vitro culture

The regenerative capacity of moss tissue, allows vegetative tissue to be propagated without difficulty in a number of ways.

Growth on agar medium

Agar medium in Petri dishes may be inoculated either with spores or with somatic tissue. For routine sub-culture using somatic tissue, it is best to use protonemal tissue from a vigorously growing culture. Chloronemal tissue, the growth of which is enhanced when ammonium is provided as nitrogen source, is easiest to sub-culture. A fragment of tissue 1 to 2 mm in diameter is sufficient to establish a new culture. Tissue other than chloronemata, *e.g.* leaf cells, may take a long time to regenerate. For growth in tubes or jars, it is important not to use lids which seal tightly.

If contamination is a problem, Petri dishes may be taped with "Micropore" medical plaster strip, without affecting the growth or development of cultures. This is a porous material that slows evaporation from the dish slightly and reduces contamination, principally because it limits air exchange during handling. Sealing cultures with paraffin film slows growth and prevents rapid regeneration. It is however satisfactory for long-term cultures which have been started with a protonemal inoculum (see below).

Growth on agar medium overlaid with cellophane

Protonemal tissue, free of agar, may be obtained by inoculating protonemal fragments onto cellophane overlaying solid medium in a Petri dish. Petri dishes containing solid minimal medium + 5mM di-ammonium tartrate are overlaid with sterilized cellophane discs (type 325P, Cannings, Avonmouth, Bristol, U.K.; the specification of the cellophane appears to be critical, but other sources may be suitable). Suspensions of protonemal fragments are prepared by blending protonemal tissue in sterile distilled water. The exact procedure will depend on the type of blender used. 500 mg fresh weight of tissue is suspended in 20 ml of water and blended until the tissue is cut into fragments containing approximately 20 to 100 cells. About 2 ml of suspension is required to inoculate a 90 mm Petri dish. After 7 days incubation in standard conditions, about 500 mg (fresh weight) of vigorously-growing protonemal tissue is obtained from a wild-type culture on a 90 mm Petri dish. Tissue may be harvested using a sterile spatula.

Growth in liquid medium

Physcomitrella can be grown in liquid culture either in shaken flasks or in a fermenter. For shaken flask culture, a tissue inoculum, prepared as for the inoculation of cellophane-overlay plates (see above), may be used to inoculate liquid minimal medium. Vigorous agitation is not necessary for growth, but growth rates in shaken liquid cultures are not as great as those obtained on Petri dishes or in a fermenter supplied with CO_2 -enriched air. Full details of fermenter culture are given in Boyd et al [6]. The maintenance of fermenter cultures, free of contamination is difficult and the production of tissue in quantities up to about 100 g is usually easier using cellophane-overlay plates.

Temperature

Physcomitrella grows on solid media in temperatures up to about 28 °C, possibly somewhat higher in liquid medium. Little difference in growth rate is observed in the temperature range 20 °C to 26 °C. Temperatures between 24 °C and 26 °C are used for routine culture. Growth is slower but still satisfactory at 15 °C and this has been used as the permissive temperature when temperature-sensitive mutants have been sought. Because of the radiant heat from the light source it is usually necessary to keep the air temperature below the desired culture temperature. The temperatures given are those of the medium on/in which cultures are grown.

Light

For routine culture, the exact quality of light provided is not critical. Most studies to date have used continuous light from fluorescent tubes at an intensity of between 5 and 20 W m^{-2}. Some laboratories use intermittent light, the most commonly used cycle being 16 h light + 8 h dark. Intermittent light results in slower development. No developmental effects of intermittent light have yet been reported, although it may allow synchronization of the cell cycle.

Media

The medium which has been used most commonly is ABC medium. However, this medium, when prepared in the way described here, has a heavy precipitate and so, notwithstanding its high calcium content, is probably variable in the amount of calcium, and possibly of other nutrients, that are available. There has therefore recently been agreement among many Physcomitrella workers to try a modified ABC medium, BCE the recipe for which is also given below.

"Standard" conditions

Throughout this chapter, standard conditions refer to growth at 25 °C, in continuous white light at intensities greater than 5 W m^{-2}.

Procedures

Culture maintenance

Cultures can be kept in a healthy state for a considerable time (at least 6 months) after they have developed to a required stage by sealing the Petri dish or test tube with paraffin film. Storage is prolonged further by keeping the sealed cultures at 15 °C. For very long term storage, moss strains can be stored in liquid nitrogen. Recovery from cryopreservation is not reliable and it is advisable to preserve a number of replica cultures. Some developmentally-abnormal mutants appear to be less tolerant of cryopreservation than developmentally-normal strains.

Cryopreservation [12]

1. Tissue of the strains to be preserved, is grown for at least 7 days under standard conditions on 90 mm cellophane-overlay plates of appropriately-supplemented minimal medium plus ammonium.
2. The tissue is transferred on the cellophane, to fresh appropriately-supplemented solid minimal medium plus ammonium plus 500 mM mannitol.
3. 1 ml of liquid medium plus ammonium plus 500 mM mannitol is added to the surface of the tissue.
4. The culture is incubated for a further seven days under standard conditions.
5. 2 ml of 5% v/v DMSO + 10% w/v glucose is added to a series of sterile Eppendorf tubes.
6. Tissue (about 1/10th of plate per tube) is added to each tube.
7. Incubate at 20 °C for 1 h.
8. Freeze the tubes at a rate of 1 °C per min to -35 °C, then place into liquid nitrogen for storage.

To retrieve cryopreserved tissue, tubes are thawed at room temperature, and the contents of each tube added to 10 ml of sterile distilled water and left for 30 min. The tissue is then inoculated onto appropriate solid medium.

Protoplast isolation and regeneration [*9, 10, 11*]

Protoplast isolation and regeneration are easy to achieve in *Physcomitrella*, and protoplasts therefore provide a way of establishing a clone from a single cell, a procedure that is often crucial for both transformation and somatic mutagenesis. Protoplasts can also be fused and so provide a method of somatic hybridization.

Protoplasts are isolated from young protonemal tissue obtained from cellophane-overlay plates.

1. Protonemal tissue is added to a sterile aqueous solution of Driselase in 8% D-mannitol (for preparation, see below), at about the rate of 100 mg of tissue (fresh weight) per ml of Driselase solution.
2. Incubate for 20 to 30 min at 25 °C with occasional gentle shaking.
3. Remove undigested tissue, by filtering sterilely through a stainless steel or nylon mesh (pore size: approx. 100 μm $\times$ 100 μm).
4. Sediment protoplasts by gentle centrifugation (100 to 200 $\times$ g for 3 min).
5. Remove supernatant carefully, and resuspend the protoplasts in about the same volume of sterile 8% (w/v) D-mannitol. Sediment as in step 4.
6. Repeat step 5 once more and finally resuspend the washed protoplast and adjust to a concentration appropriate to the procedure involved. About 10^6 viable protoplasts are obtained from 1 g fresh weight of tissue.

Protoplasts are usually regenerated embedded in soft agar medium overlaying cellophane, in turn overlaying osmotically-buffered medium. Once regenerated, the cellophane and protoplast can be transferred to medium without mannitol, where development is more rapid.

7. 1 volume of protoplast suspended in 8% D-mannitol solution is added gently to 10 volumes of molten PRT medium (see below) which has been maintained at 42 °C.
8. The mixture is pipetted or poured gently but quickly, onto cellophane overlaying PRB medium. 1 ml of top layer covers a 90 mm Petri dish, 400 μl covers a 50 mm dish.
9. After incubation under standard conditions for 3 to 4 days, the protoplasts will have regenerated cell walls and may be transferred to fresh medium without mannitol. Regeneration of *Physcomitrella* protoplasts requires a high light intensity ($>$ 5 Wm^{-2}).

Mutagenesis

N-methyl-N'-nitro-N-nitrosoguanidine (NTG) mutagenesis

N-methyl-N'-nitro-N-nitrosoguanidine is a very potent mutagen and should only be used by those who have been trained in mutagen handling. Since it is a powder, extreme caution must be taken when weighing this compound and local safety rules for mutagen handling must be observed.

Using spores [1]

1. Prepare 10 ml of a spore suspension containing about 10^5 spores per ml in sterile Tris-maleate buffer (pH 6).
2. Dissolve 1 mg of NTG in a separate 10 ml of Tris-maleate buffer (pH 6) and incubated for 30 min at 25 °C.
3. Mix the spore suspension with the NTG solution.
4. Incubate for 30 min at 25 °C, shaking gently from time to time.
5. Centrifuge at 2500 × g for 5 min Discard the supernatant.
6. Resuspend the spores in 20 ml distilled water.
7. Repeat steps 5 and 6 twice more.

The spores are now ready for plating. About 10% of the originally viable spores survive this treatment. The mutagenic treatment results in an initial delay in development but once germinated, the sporelings grow as usual.

Using somatic tissue [5]

The method used for somatic mutagenesis is identical to that employed with spores except that the NTG solution is added to 1 g (fresh weight) of young protonemata obtained by growth on cellophane-overlay plates which has been pre-incubated for 30 min at 25 °C in 10 ml of Tris-maleate buffer. The mixture is then incubated for a further 60 min at 25 °C with gentle agitation. The treatment is terminated by filtering the mixture sterilely using a stainless steel or nylon mesh (pore size: approximately 100 μm × 100 μm) which retains the protonemal tissue and by washing thoroughly, with sterile, distilled water.

This procedure results in a cell survival rate of about 10%. If individual cells are required these may be obtained by protoplasting the tissue using the method described above.

UV mutagenesis

Although not as an effective mutagen as NTG, UV irradiation is less hazardous if used with the correct safety precautions. The mutagenic UV source will need to be standardized before use. Most studies aim for spore/cell survival around 5%.

Using spores

20 ml of a spore suspension in water is irradiated in an open 90 mm Petri dish. The spore suspension must be kept in darkness for 24 h following UV irradiation to limit repair.

Using somatic tissue

Seven-day old tissue growing on cellophane-overlay plates can be irradiated directly with UV. Tissue must be held in darkness for 24 h following UV irradiation to limit repair. If individual cells are required, the tissue can be protoplasted following the dark incubation.

Mutant isolation from either spores or protoplasts

Auxotrophic mutants [1, 5]

No satisfactory selective procedure for obtaining auxotrophic mutants is yet available. All such mutants have been obtained so far by non-selective isolation. This procedure involves culturing protonemata derived from mutagenized spores or protoplasts, initially on medium containing the substances which are required for the growth of the classes of nutritionally-deficient mutant which are being sought. After 1–2 weeks' growth on this medium, the protonemata can be tested for auxotrophies by transferring half of each culture on to supplemented medium and the other half on to minimal medium. 25 protonemata can be tested in this way in pairs of 9 cm petri dishes. In order to isolate a particular kind of auxotroph, it may be necessary to test several thousands of somatic clones, each of which has grown from a single mutagenized spore or protoplast. Analogue-resistant strains could also be obtained using total isolation but it is easier to isolate them selectively by including the analogue in the growth medium either from the outset, as for example for resistance to D-serine or p-fluorophenylalanine, or as soon as mutagen-treated spores/protoplasts have germinated/begun to regenerate, as for resistance to 8-azaguanine.

Isolation of developmentally-abnormal strains [2, 3, 8]

The most straightforward way to isolate a wide range of morphologi-
cally-altered mutants is to inoculate mutagenized spores or protoplasts at a density of about 100 survivors per 9 cm petri dish and incubate them for 3 to 4 weeks under standard conditions. At this time,
abnormal strains blocked or altered in various stages of gametophytic
development will be readily observable and can be saved for further
study by subculturing on to fresh medium.

Procedures for genetic analysis

Sexual crossing [1, 2, 4]

Gamete production in *Physcomitrella* requires a temperature below 18 °C. Since *Physcomitrella* is monoecious, it is self-fertile but crossing can be ensured by the inclusion of auxotrophies, *eg* requirements for nicotinic acid or p-amino benzoic acid, in the strains to be crossed. Such auxotrophic strains are cross-fertile, but self-sterile on normally-supplemented medium [7]. The two strains to be crossed are inoculated near to each other, on minimal medium. After about 3 weeks growth in standard conditions, the culture is transferred to a temperature of 15 °C to 18 °C. After a further 3 weeks, sterile water is added to the culture to allow the male gametes to swim to the female gametes and effect fertilization. Sufficient water is added to make the culture thoroughly damp but not submerged. Sporophytes should be visible within a further two weeks, but if none are formed, further irrigation may be necessary. The timings given here are relaxed, and successful crosses have been made using shorter intervals. It is probable that the protocol for crossing could be improved.

Somatic hybridization [4, 8, 9]

Many developmentally-abnormal strains are sexually sterile, but can be analysed genetically by the use of somatic hybrids, which can be used to test for dominance of mutant phenotypes and for complementation between two phenotypically-similar developmental mutants [11]. Somatic hybrids involving recessive characters are fertile and can be used for further genetic analysis.

The method using polyethyleneglycol (PEG) -induced protoplast fusion, to obtain somatic hybrids, first described by Grimsley et al [9, 10], is still used widely. Electrofusion of protoplast has also been used [18]. Either method relies on the use of complementing auxotrophies in the strains to be hybridized to allow the selection of the hybrid.

Transformation

Transformation of *Physcomitrella* can be carried out using plasmid DNA, containing a gene coding for resistance to either G418, or hygromycin or sulfadiazine. Selection for transformants is made using G418 at 50 μg/ml, hygromycin at 30 μg/ml and sulfadiazine at 150 μg/ml. Following transformation three classes of antibiotic-resistant regenerant are obtained:

transient: Do not retain resistance upon sub-culture.
unstable: Grow slowly on selective medium. Resistance lost when selection is relaxed. Probably not transmitted through meiosis.
stable: Grow on selective medium almost as fast as on non-selective medium. Resistance retained when selection is absent. Transmitted in a regular Mendelian manner through meiosis.

Transformation procedures are still being refined, but pBR-derived vectors appear to be most reliable, while the physical state of the DNA does not seem to be a significant factor influencing transformation rates. Two methods for transformation are used routinely. PEG-mediated DNA uptake by protoplasts is reliable and requires no special apparatus but micro-projectile bombardment gives higher transformation rates and this method can probably be improved further.

Transformation using PEG-mediated DNA uptake by protoplasts [17]

1. Protoplasts are prepared as described above.
2. After the second wash, estimate protoplast density using a haemocytometer. Centrifuge (100 to 200 × g for 3 min) and resuspend in sufficient D-mannitol/MgCl$_2$/MES solution to give a final protoplast density of 1.6×10^6/ml.
3. Meanwhile, prepare DNA to be used in transformation by dispensing 20 to 50 μg of DNA dissolved in no more than 30 μl TE, into sterile 10 ml tubes. Centrifuge gently to bring DNA solution to bottom of tube.
4. Add 300 μl protoplast suspension from step 2 to DNA from step 3.
5. Add 300 μl PEG/T solution.
6. Heat for 5 min at 45 °C. Return to room temperature (20 °C) for 10 min.
7. Add 1 ml 8% D-mannitol solution. Invert gently to mix. Wait 1 min.
8. Add 2 ml 8% D-mannitol solution. Invert gently to mix. Wait 1 min.
9. Add 7 ml 8% D-mannitol solution. Invert gently to mix. Wait 1 min.
10. Centrifuge (100–200 × g for 3 min). Remove most of supernatant, retaining about 500 μl, gently resuspend protoplast in the residual supernatant.
11. Add protoplasts to 10 ml molten PRT medium (at 42 °C). Pour immediately onto 3 Petri dishes containing PRB medium overlaid with cellophane.
12. Incubate in standard conditions for 3 days
13. Transfer regenerating protoplasts on top layer to appropriate selective medium.

Transformation using micro-projectile bombardment

This method was developed using a gun to the design of Lonsdale *et al.* [14], which uses polycarbonate macroprojectiles and 0.22 inch calibre blank launcher cartridges. This design gives best results with a distance of 150 mm between the stopper plate and the tissue, and with two discharges per petri dish of tissue (the position of the dish is moved between discharges).

1. Suspend 50 mg of dry tungsten powder (M17 grade, Sylvania, Towanda, PA 18848, USA) in 300 μl of absolute ethanol, mix vigorously, and centrifuge at 10,000 $\times$ g for 5 min. Withdraw the ethanol carefully.

2. Wash the tungsten by the addition of 1.5 ml of sterile distilled water. Resuspend and centrifuge. Repeat twice more and finally resuspend in 1 ml of 50% (w/v) sterile glycerol.

3. To 25 μl of the tungsten suspension add in turn:

       5 μl of a solution of 1 μg plasmid DNA/μl TE

       25 μl sterile 2.5 M CaCl$_2$ solution

       10 μl sterile 0.1 M spermidine (free base) solution

   Mix gently and allow to stand for 10 min. Carefully withdraw 35 μl of the supernatant and discard. Transfer to ice until used for bombardment.

4. The addition of CaCl$_2$ and spermidine causes DNA-tungsten complexes to from. These must be dispersed immediately before use. Flicking the tube with the finger is effective. Immediately after dispersal, place 3 μl of DNA-tungsten complex on the macroprojectile (or as appropriate for the method of micro-projectile bombardment used).

5. Bombard 6 to 7 day old tissue grown on cellophane-overlay plates containing minimal medium with ammonium as nitrogen source. Remove the lid from the Petri dish and cover with sterile stainless steel mesh (aperture size 1 mm $\times$ 1 mm). Evacuate the chamber to 28 mbar before discharge.

6. Incubate the tissue under standard conditions for 48 h.

7. After 48 h incubation, the tissue may be transferred on the cellophane directly to selective medium, or may instead be harvested and either blended and plated onto selective medium (as described above for the preparation of cultures on cellophane-overlay plates), or, if single-cell clones are required, protoplasted before plating onto osmotically-buffered selective medium.

Media, supplements and solutions [13]

Unless stated otherwise, sterilization is by autoclaving at 121 °C for 20 min.
Analytical grade inorganic chemicals should be used where possible.

Stock solutions for growth media

These are best dispensed into convenient aliquots and stored frozen. Under these
conditions there is no need to sterilize these solutions before use.

Solution A

$Ca(NO_3)_2.4H_2O$	118 g
$FeSO_4.7H_2O$	1.25 g
distilled H_2O	to 1 l

Solution B

$MgSO_4.7H_2O$	25 g
(**or** *anhydrous MgSO₄*	12 g)
distilled H_2O	to 1 l

Solution C

KH_2PO_4	25 g
distilled H_2O	500 ml

Adjust pH to 6.5 with minimal volume of 4 N KOH and make up to 1 l with
additional distilled H_2O.

Solution E

KNO_3	101 g
Ferric citrate.H_2O	263 mg
distilled H_2O	to 1 l

Hoagland's A-Z trace element solution (TES):

H_3BO_3	614 mg	$MnCl_2.4H_2O$	389 mg
$Al_2(SO_4)_3.K_2SO_4.24H_2O$	55 mg	$CoCl_2.6H_2O$	55 mg
$CuSO_4.5H_2O$	55 mg	$ZnSO_4.7H_2O$	55 mg
KBr	28 mg	KI	28 mg
LiCl	28 mg	$SnCl_2.2H_2O$	28 mg
		distilled H_2O	to 1 l

Hoagland's TES is widely used but the exact composition of the TES is probably not important.

Growth supplements

substance	concentration in the medium	weight per litre
adenine	500 μM	67.5 mg
p-aminobenzoic acid	1.8 μM	247 μg
di-ammonium (+) tartrate	5 mM	920 mg
nicotinic acid	8 μM	1 mg
D-sucrose	15 mM	5 g
thiamine HCl	1.5 μM	0.5 mg

All the above supplements, with the exception of adenine, may be kept as aqueous 100 × concentrated stock solutions, sterilized by autoclaving, and added to growth media as required to give the concentrations listed above. Adenine is best added as a solid, and medium containing adenine may be autoclaved.

Growth media

Agar

It is probable that any high grade agar, such as Sigma High Gel Strength Agar (cat. # A9799), can be used to gel moss media. The quantities in the recipes below refer to Sigma Agar #A9799, and may need to be altered for other agars. Either heat media containing agar, to dissolve agar before aliquoting and autoclaving, or add the appropriate amount of agar to each individual aliquot, autoclave and disperse/dissolve the agar before medium solidifies.

ABC minimal medium

This is a modified Knop's medium that contains a high level of calcium and which precipitates on autoclaving. Although ABC medium has been used in many past studies, BCE medium may be give more reproducible results.

solution A	10 ml
solution B	10 ml
solution C	10 ml
TES (either)	1 ml
(agar	8g)
distilled H_2O	to 1 l

The pH of the autoclaved medium will be between 5.3 and 5.9, the actual value depending upon the type of agar used, and is not usually adjusted.

BCE minimal medium

This medium contains no calcium, which must be added (as $CaCl_2$) as required. 1mM calcium is now being assessed for routine use, but note that protoplast regeneration requires at least 5mM calcium.

solution B	10 ml
solution C	10 ml
solution E	10 ml
TES (either)	1 ml
(agar	8 g)
distilled H_2O	to 1 l

PRB (protoplast regeneration medium, bottom layer)

liquid BCE medium	1 l	
$CaCl_2.6H_2O$	2.19 g	(= 10mM)
D-mannitol	60 g	
di-ammonium(+)tartrate	920 mg	(=5mM)
agar	8 g	

PRT (protoplast regeneration medium, top layer)

liquid BCE medium	1 l	
$CaCl_2.6H_2O$	2.19 g	(= 10mM)
D-mannitol	80 g	
di-ammonium(+)tartrate	920 mg	(=5mM)
agar	8 g	

Other solutions.

1M $Ca(NO_3)_2$

$Ca(NO_3)_2.4 H_2O$	236.1 g
distilled water	1 l

Sterilize by autoclaving. Store at 4 °C.

Driselase solution

Driselase	1 to 2 g (depending on batch)
D-mannitol	8 g
distilled H_2O	to 100 ml

1. Stir to mix but do not shake vigorously.
2. Leave to stand at room temperature for 15 min.

3. Centrifuge at $2500 \times g$ for 5 min.
4. Remove the clear supernatant and filter sterilize.

D-mannitol/Ca(NO₃)₂ solution

8% (w/v) D-mannitol solution	9 ml
1M $Ca(NO_3)_2$ solution	1 ml
1M Tris buffer, pH 8.0	100 ml

Make up fresh, on day of use. Filter sterilize

D-mannitol/MgCl₂/MES solution

D-mannitol	9.1 g
distilled water	8.85 ml

Sterilize by autoclaving and store at room temperature. On day of use, add:

1M $MgCl_2$ (203.3 g $MgCl_2.6H_2O/$ l)	150 μl
1% MES pH5.6 solution	1ml

Filter sterilize

1% MES pH 5.6

Use 1% (w/v) 2-[N-morpholino]ethanesulphonic acid in distilled water. Adjust to pH 5.6 with 0.1M KOH. Sterilize by autoclaving. Store at 4 °C.

PEG/T (PEG solution for transformation)

1. Autoclave 2 g PEG 6000 in a glass container.
2. On day of transformation, melt PEG in microwave.
3. Add 5 ml D-mannitol/Ca(NO₃)₂ solution and mix well.
4. Leave at room temperature for 2 to 3 h before use

TE

Tris-(hydroxymethyl)-aminomethane	1.21 g
diaminoethanetetra-acetic acid	372 mg
distilled water	1 l

Adjust to pH 8.0 with 0.1M HCl. Sterilize and store at 4 °C.

1M Tris buffer (pH 8.0)

Tris-(hydroxymethyl)-aminomethane	121.1 g
distilled water	1 l

Adjust to pH 8.0 with 0.1M HCl. Sterilize and store at 4 °C.

Tris-maleate buffer (pH 6)

Tris-(hydroxymethyl)-aminomethane	6 g
maleic acid	6 g
distilled H$_2$O	to 1 l

Adjust pH to 6.0 with 10M NaOH or KOH. Store at 4 °C.

References

1. Ashton NW, Cove DJ (1977) The isolation and preliminary characterisation of auxotrophic and analogue resistant mutants of the moss, *Physcomitrella patens*. Molec Gen Genet 154: 87–95.
2. Ashton NW, Cove DJ, Featherstone DR (1979) The isolation and physiological analysis of mutants of the moss, *Physcomitrella patens*, which over-produce gametophores. Planta 144: 437–442.
3. Ashton NW, Grimsley NH, Cove DJ (1979) Analysis of gametophytic development in the moss, *Physcomitrella patens*, using auxin and cytokinin resistant mutants. Planta 144: 427–435.
4. Ashton NW, Boyd PJ, Cove DJ, Knight CD (1988) Genetic analysis in *Physcomitrella patens*. In: Glime JM (Ed.) Methods in Bryology, pp 59–72, Hattori Botanical Laboratory, Nichinan, Japan.
5. Boyd PJ, Grimsley NH, Cove DJ (1988) Somatic mutagenesis of the moss, *Physcomitrella patens*. Molec Gen Genet 211: 545–546.
6. Boyd PJ, Hall J, Cove DJ (1988). An airlift fermenter for the culture of the moss *Physcomitrella patens*. In: Glime JM (Ed.) Methods in Bryology, pp 41–45, Hattori Botanical Laboratory, Nichinan, Japan.
7. Courtice GRM, Ashton NW, Cove DJ (1978) Evidence for the restricted passage of metabolites into the sporophyte of the moss, *Physcomitrella patens*. J Bryol 10: 191–198.
8. Courtice GRM, Cove DJ (1983) Mutants of the moss, *Physcomitrella patens* which produce leaves of altered morphology. J Bryol 12: 595–609.
9. Grimsley NH, Ashton NW, Cove DJ (1977) The production of somatic hybrids by protoplast fusion in the moss, *Physcomitrella patens*. Molec Gen Genet 154: 97–100.
10. Grimsley, NH, Ashton NW, Cove DJ (1977) Complementation analysis of auxotrophic mutants of the moss, *Physcomitrella patens* using protoplast fusion. Molec Gen Genet 155: 103–107.
11. Grimsley, NH, Featherstone DR, Courtice GRM, Ashton NW, Cove DJ (1979) Somatic hybridization following protoplast fusion as a tool for the analysis of development in the moss, *Physcomitrella patens*. In: Advances in Protoplast Research, pp 363–376. Proceedings of the 5th International Protoplast Symposium, Szeged, Hungary. Academiai Kiado, Budapest, Hungary.
12. Grimsley NH, Withers LA (1983) Cryopreservation of cultures of the moss *Physcomitrella patens*. Cryoletters 4: 251–258.
13. Knight CD, Cove DJ, Boyd PJ, Ashton NW (1988) The isolation of biochemical and developmental mutants in *Physcomitrella patens*. In: Glime JM (Ed.) Methods in Bryology, pp 47–58, Hattori Botanical Laboratory, Nichinan, Japan.
14. Lonsdale D, Onde S, Cuming A (1990) Transient expression of exogenous DNA in intact, viable wheat embryos following particle bombardment. J Exp Bot 41: 1161–1165.
15. Sawahel W (1994) Genetic transformation of the moss, *Physcomitrella patens*. PhD thesis, University of Leeds, U.K.
16. Sawahel W, Onde S, Knight CD, Cove DJ (1992) Transfer of foreign DNA into Physcomitrella patens protonemal tissue by using the gene gun. Plant Mol Biol Rep 10: 315–316.
17. Schaefer D, Zryd J-P, Knight CD, Cove DJ (1991) Stable transformation of the moss *Physcomitrella patens*. Molec Gen Genet 226: 418–424.
18. Watts JW, Doonan JH, Cove DJ, King JM (1985) Production of somatic hybrids of moss by electrofusion. Molec Gen Genet 199: 349–351.
19. Wettstein, Fr von (1932) Genetik. In: Vedoorn F (Ed.) Manual of Bryology, pp 233–272, Martinus Nijhoff, The Hague, Netherlands.

Plant Tissue Culture Manual **H4**, 1–25, 1996.
© 1996 *Kluwer Academic Publishers.*

Thin cell layer (TCL) method to programme morphogenetic patterns

K. TRAN THANH VAN & C. GENDY
Institut de Biotechnologie des Plantes, CNRS Université Paris-Sud, 91405 Orsay, France

Introduction

The establishment of plant organ shape, structure and function proceeds from the embryo stage to the adult stage according to an ordered sequence of events. This implies that the genetic information is expressed or repressed according to a temporal and spatial order.

Due to the fact that growth and development events can be changed, to a certain extent, by a large array of factors which apparently lack specificity, the analysis at the biochemical molecular level cannot be easily conducted especially when developmental mutants are not available, as is the case for most plant species.

As an alternative to the study of developmental mutants, it would be valuable to be able to "programme" different morphogenetic patterns experimentally.

To date, the tobacco Thin Cell Layer (TCL) experimental system, consisting of a few cell layers (the epidermal layer and 3 to 6 cortical layers of differentiated cells) with reduced sizes (1mm × 5 or 10mm), is a unique system in which not only all patterns of morphogenesis known in plants can be programmed directly (without an intermediate callus phase), but also patterns which are new for the species studied. For example, in tobacco, somatic proembryos can be induced from subepidermal cells [14, 15, 16, 17, 19].

Markers of morphogenetic differentiation can be delineated in experimental systems in which different patterns can be : i) followed at the cell level from the early stage, ii) controlled by specific molecules, iii) inhibited by inhibitors of specific metabolic pathways and iv) recovered upon addition of such compounds. The identification of cytological and biochemical markers can lead to the identification of the gene(s) involved in specific steps of morphogenetic differentiation and to the study of their function(s) [3, 4, 7, 9, 12, 18, 20].

Application of the TCL method to recalcitrant species has allowed us and other researchers to overcome the difficulty in obtaining regeneration. For the study of gene function as well as for the production of transgenic plants, especially crop plants, for improvement purposes, the control of regeneration constitutes, at the present time, a serious limitation in plant biotechnology despite the availability of refined modern techniques of analysis at the molecular level.

The TCL method, originally developed on *Nicotiana tabacum* has been successfully applied and/or adapted to other species. Longitudinal TCLs are used in order to allow the analysis of mechanisms of cell differentiation and organogenesis from a defined cell layer whereas transverse TCL (tTCL) is used when the first objective is to overcome the difficulty in obtaining organ regeneration and/or somatic embryogenesis. As both longitudinal and transverse TCLs consist of one or a few cell layers, the content of their endogenous factors which are most of the time unknown, is assumed to be minimal, compared to the exogenous conditions applied in which the cells are cultured. Therefore, they are more medium-dependent than the classical type of explant which is larger and more complex.

In summary, the advantages of TCLs are as follows : i) when comprising only differentiated cells (epidermal and cortical cells), as in the case of longitudinal TCLs, the TCL method allows the analysis of the programmed patterns from the early stages to the full development of the organ, and therefore allows its study at the temporal level; ii) the minimal size of TCL and the well defined cell layer from which morphogenetic events start facilitate this study at a spatial level; iii) the rapidity (12, 14 days) in obtaining different patterns of morphogenesis including fully developed flowers has led to a reduced time in obtaining transgenic plants and transgenic seeds *in vitro* [2]; iv) the high number of organs obtained per TCL explant (up to 50 flowers or 800 vegetative buds) has also led to the high frequency in the production of transgenic plants or seeds.

To date, our group has succeeded, by using TCL methods, in reprogramming morphogenesis in several species including species of *Nicotiana, Torenia, Petunia, Brassica, Nautilocalyx, Saintpaulia, Begonia, Arabidopsis, Soja biloxi, Vicia faba, Psophocarpus*. For recalcitrant species such as *Phalaenopsis* and crop plants as well as for non-recalcitrant species with small sized organs (such as in *Arabidopsis*), one can use TCL to control: i) organogenesis in *Arabidopsis*, ii) both organogenesis and somatic embryogenesis in wheat, *Sorghum* (in collaboration with M. Sené, L. Bui, J. Vidal and P. Gadal), iris (in collaboration with H. Schricke and D. Joulain) and iii) somatic embryogenesis in other monocotyledons (bamboo [5]), and *Digitaria* (in collaboration with L. Bui, T. Do, J. Vidal and P. Gadal).

Other groups have successfully applied TCL methods to *Nicotiana tabacum* [1, 11], *Helianthus* [10] and *Pisum sativum* [6] in order to obtain programmed patterns including flower development, somatic embryogenesis and regeneration of vegetative buds respectively.

Procedures

I - Longitudinal TLC method

Model system: *Nicotiana tabacum* cv Samsun, Wisconsin and Xanthi.

1. Culture conditions of the donor plants:

Steps in the procedure
1. Grow plants from seeds in pots (10 cm diameter) in vermiculite.
2. Water with a mineral solution three times a week and on every other days, with deionized water (for both types of watering, 700 ml per day).
   Mineral solution:
   (in mM) 2.71 KNO_3, 1.00 KH_2PO_4, 1.11 $MgSO_4$, 1.04 $(NH_4)_2 SO_4$, 4.64 $Ca(NO_3)_2$,
   (in μM) 110.00 Na_2 Fe-EDTA, 36.75 KCl, 48.52 $H_3 BO_3$, 10.06 $MnSO_4$, 0.95 $ZnSO_4$, 0.55 $CuSO_4$, 0.22 $(NH_4)_6 Mo_7O_2$
   and (in nM) 2.55 H_2SO_4.
3. The environmental conditions are : 24 °C $\pm$ 2 °C for temperature, 65% relative humidity, 16h photoperiod of natural light complemented by artificial irradiation of 25 W. m^{-2} when the natural irradiance decreases below 50 W. m^{-2}
4. After three months, the plants reach the floral stage. When the terminal flower of the inflorescence becomes a green fruit and the donor plant bearing 5 – 80 green fruits, the physiological stage of the plant is appropriate for TCL.

2. Preparation of TCL

A. Sampling

Steps in the procedure
1. At this precise physiological stage, select 2–3 floral branches of 5 to 10 cm of length below the top of the inflorescence from each donor plant. Three to four donor plants i.e. 6 to a 12 floral branches in total are used.
2. Cut the floral branches from the inflorescence.
3. Wash in water, then in water with teepol and surface sterilize with 7% of calcium hypochlorite for 10 minutes.
4. Rinse thoroughly (five times) in sterile water and store in sterile water for the next phases of the procedure.

B. Excision of TCL

Steps in the procedure

1. Excise several long ribbons of TCL (1mm width and 4 to 8 cm length from each floral branch; Fig 1), using special microscalpels

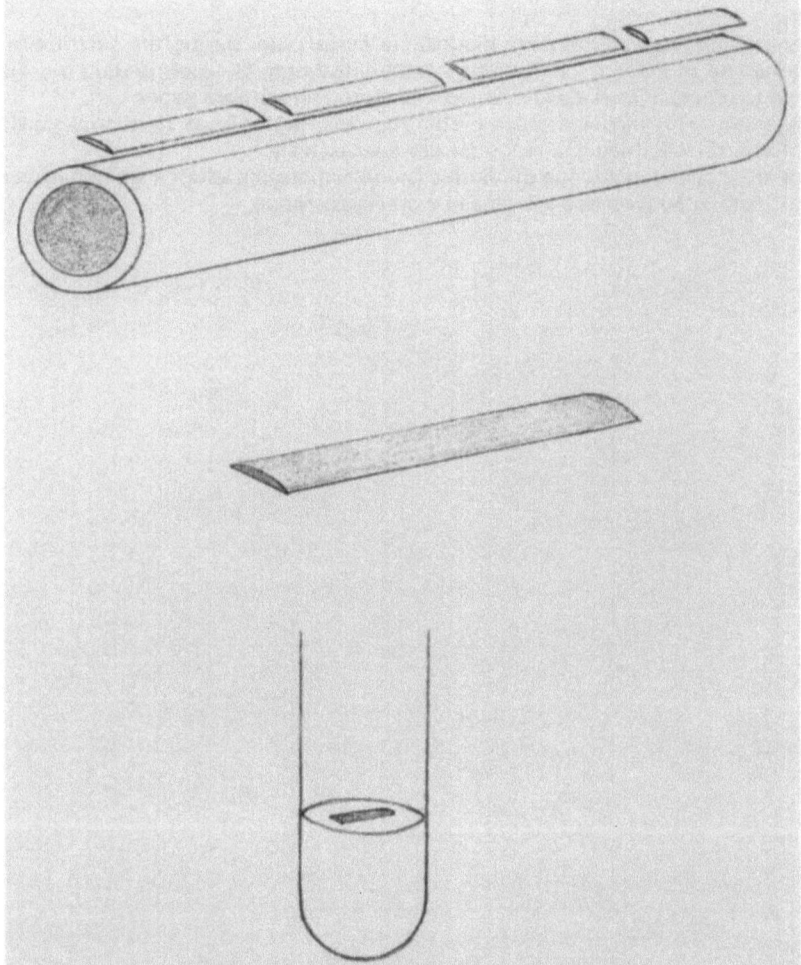

Fig. 1. Schematic of floral branches and excision of TCL.

made in the laboratory : triangular pieces (7mm × 2mm) of razor blade mounted on special metal handles of the type used in micro-surgery (supplier: Moria).

2. Put these ribbons in a 10 cm diameter Petri dish and cover them with a film of sterile water.

3. Cut TCLs of 6 to 8 mm in length from each TCL ribbon in the Petri dish or directly on the floral branch (Fig 1). A pool of 20–25 TCLs from different floral branches is made and kept under a fine film of water in Petri-dishes.
4. Inoculate TCLs randomly onto a solidified medium and randomly distribute them to different types of culture medium.

Notes
– Only the handles (not the razor blades) are autoclaved. During the excision of TCL ribbons and of TCLs, microscalpels are sterilized only by alcohol, then thoroughly rinsed (5 times) in sterile water, dried and stored under filter paper.
– Only sharp razor blades are used. This requires that at least 10 microscalpels are sterilised, rinsed, dried and ready for use successively.
– TCLs are inoculated into the medium as soon as possible when a small number e.g. 20 to 25, (5 to 10 for less experienced worker) are made.

3. Inoculation of TCL

Steps in the procedure

1. Only a TCL method using solidified medium is described here. Use test tubes (15 cm length, 2.5 cm diameter) as containers for medium solidified with gelose Difco (10g/l) or gelrite (3g/l) except for flower programme.
2. Pour 25 ml of culture medium into each test tube.
3. Use forceps to pick up gently TCL (stored in Petri-dishes) and place them on the surface of the culture medium, **the epidermis side upwards**.

Notes
- Use only fine forceps with tips of 1 or 2 mm wide. Select forceps for which the distance between the tips does not exceed 1 cm, so as to reduce the pressure exerted on the TCL while manipulating it.
- Test tubes are covered by a metal lid (Bellco type) which allows gas exchange through a cotton wool ball placed at the bottom of the lid. Hermetic types of closure should be avoided. In order to ensure that the TCL is closely applied on the surface of the culture medium (but not immersed in the medium), a slight pressure is made on the TCL, the two tips touching the two poles of the TCL.
- Container shape and volume as well as the type of sealing, and so gas exchange, are important factors for "flower programme". Therefore, Petri dishes of 1 cm height, and sealing with parafilm, should be avoided.
- Multiwell (1 cm diameter) plates of 20 wells (supplier: Falcon) can be used, and so the volume of the medium can be reduced to 1.5 ml.

4. Culture conditions for TCLs

Once inoculated in test tubes, incubate in a growth chamber at 24 °C $\pm$ 1 °C under an irradiance of 20 W. m^{-2}. The TCL should receive the light vertically and not laterally. The relative humidity is 70%, the photoperiod can be 16h or 24 h. For root programme, the TCLs must be kept in the dark when solified medium is used.

5. Culture medium

A. Composition:
the culture medium contains macro and microelements of Murashige and Skoog [8], and myo-inositol (100 mg/l). IBA, kinetin and glucose are added at different concentrations (Table 1) to induce separately different morphogenetic programmes (Fig. 2). The medium pH is adjusted to 5.6. Gelose is added at 10 g/l, gelrite at 3g/l.

Table 1. Glucose, IBA and Kinetin concentrations for different morphogenetic programmes on tobacco thin cell layer (TCL).

Morphogenetic Programme	Glucose (g/L)	IBA (molar)	Kinetin (molar)
Flower	30	10^{-6}	10^{-6}
Vegetative Bud	30	10^{-6}	10^{-5}
Root	10	10^{-5}	10^{-7}
Callus	30	3.10^{-6}	10^{-7}

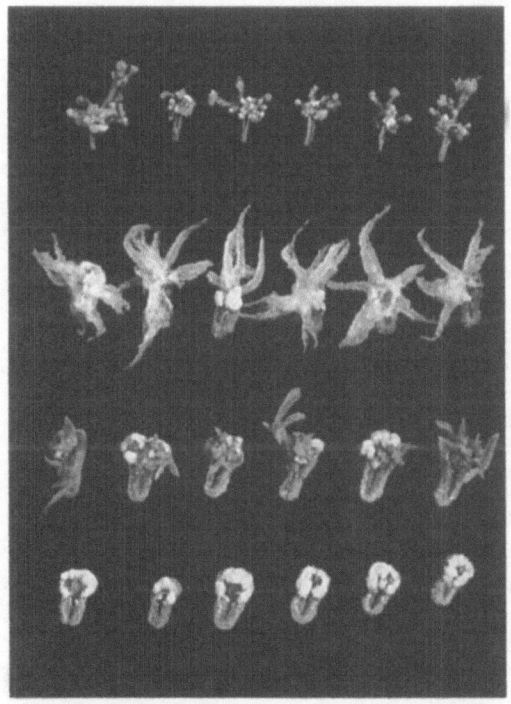

Fig. 2. Four morphogenetic programmes from tobacco TCL (stage: 14 days).
From the top : first line : Flower Programme (flowers are formed directly on the surface of the TCL)
Second line : Root Programme
Third line : Vegetative buds Programme
Fourth line : Callus Programme

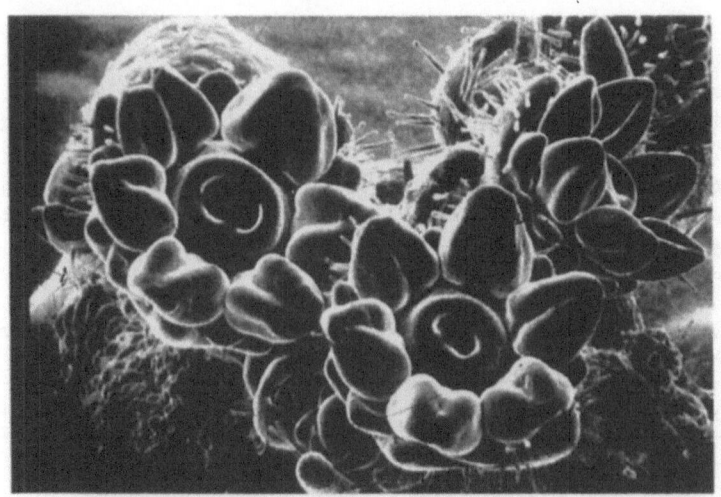

Fig. 3. Young flowers (approximately 50 in total) formed directly on one tobacco TCL (stage: 14 days).

Fig. 4. Fully developed flower from Petunia TCL (stage : 4 weeks).

Functional flowers are obtained within 12–14 days (Fig. 3). Fully developed flowers are obtained within 3 weeks for tobacco TCL, 4–5 weeks for Petunia TCL (Fig. 4).

B. Autoclaving:
The medium is autoclaved at 115 °C for 20 minutes under 1 bar pressure.

6. Observations

Morphogenetic changes are scored every two days (starting from day 4 to day 28 after the beginning of the culture) by observation under a stereo microscope and if necessary scanned with a scanning electron microscope.

The number of organs formed (flowers, vegetative buds, roots), per TCL and the number of the TCLs responding to the treatment are analysed statistically.

II – Transverse TCL (tTCL) method

Model system : *Iris pallida, Phalaenopsis, Oryza sativa, Sorghum, Digitaria, Arabidopsis, Panax ginseng*

This method is applied to recalcitrant species (monocotyledons, woody species, legumes) and non-recalcitrant species of small size.

Steps in the procedure
1. Grow donor plants in appropriate environmental conditions (see above, section I.1).
2. Select the optimal size for tTCL method, by making transverse sections of various sizes [e.g. from 1 mm, 1.5 to 2 mm, 500 μm–1 mm or 300–200 μm (using a vibratome)]. All plant organs can be used: root, mesocotyl, hypocotyl, epicotyl, cotyledon of immature/mature/germinating embryos, petiole, leaf blade, leaf vein, internode, node, stem, immature inflorescence, inflorescence, filament of anther, anther, carpel, ovary style, petal, sepal, etc.
3. Inoculate the tTCL into Petri dishes (10 cm diameter) or multiwell plates, on the surface of media solidified by gelose (10 g/l). Gelrite (2 g/l) or agarose (8 g/l) are used for monocotyledons (e.g. *Sorghum*, wheat, barley, rice, *Phalaenopsis*).
4. Inoculate the tTCL in rows of different sizes. For the first phase of the investigation, the proximal or distal poles can be put at random on the surface of the medium. However, the order of each tTCL on the original organ should be respected, in order to detect the influence of the gradient.
5. From this phase of the study, observations can be made on i) the influence of the polarity and ii) the optimal size of the tTCL to programme morphogenesis. If polarity between the basal and proximal poles is shown (by the difference in the responses obtained when inoculated at random), either reduce the size of the tTCL or consider the polarity for the next phase of the study. Tissues can be arranged in rows with the proximal pole placed on the medium, or rows with the distal pole, on the medium.
6. Other parameters such as cell proliferation, cell expansion, organogenesis and/or embryogenesis can be scored between 4 to 28 days.

Acknowledgments

We thank all scientists of our group, especially M. Drira, Dien N., H. Chlyah, A. Chlyah-Arnason, A. Cousson, F. Jullien, H. Trinh, M. Mulin, V. Marfa, P. Toubart, L. Richard, Van Lê Bui, In Ok Ahn and Thao Do and of other groups who have contributed to the development of the TCL method. Yan Norry and C. Jehanno are deeply acknowledged for the illustrations of this chapter.

References

1. Altamura MM, Capitani F, Serafini-Fracassini D, Torrigiani P, Falasca G (1991) Root histogenesis from tobacco thin cell layers. Protoplasma 161: 31–42.
2. Ammirato V (1987) Speeding transgenic plants. Bio/Technology 5: 1015.
3. Cousson A, Trân Thanh Vân K (1993) Influence of ionic composition of the culture medium on *de novo* flower formation in tobacco thin cell layers. Can. J. Bot. 71: 506–511.
4. Kay LE, Basile DV (1987) Specific peroxidase isoenzymes are correlated with organogenesis. Plant Physiol. 84: 99–105.
5. Jullien F, Trân Thanh Vân K (1994) Micropropagation and embryoid formation from young leaves of *Bambusa glaucescens* "Golden Goddess". Plant Science 98: 199–207.
6. Jullien F, Wyndaele R (1992) Precocious *in vitro* flowering of soybean cotyledonary nodes. J. Plant Physiol. 140: 251–253.
7. Meeks-Wagner D, Dennis E, Tran Thanh Van K, Peacock W (1989) Tobacco genes expressed during *in vitro* floral initiation and their expression during normal plant development. Plant Cell 1: 25–35.
8. Murashige T, Skoog F (1962) A revised medium for a rapid growth and bioassay with tobacco tissue culture. Physiol. Plant. 15: 473–479.
9. Neale A, Wahlleithner J, Lund M, Bonnet H, Kelly A, Meeks-Wagner D, Peacock W, Dennis E (1990) Chitinase, β-1, 3-Glucanase, osmotin, and extensin are expressed in tobacco explants during flower formation. Plant Cell 2: 673–684.
10. Pélissier B, Bouchefra O, Popin R, Freyssinet G (1990) Production of isolated somatic embryos from sunflower thin cell layers. Plant Cell Reports 9: 47–50.
11. Rajeevan MS, Lang A (1993) Flower-bud formation in explants of photoperiodic and day-neutral *Nicotiana* biotypes and its bearing on the regulation of flower formation. Proc. Natl. Acad. Sci. USA 90: 4636–4640.
12. Richard L, Arro M, Hoebecke J, Meeks-Wagner DR, Trân Thanh Vân K (1992) Immunological evidence of thaumatin-like proteins during tobacco floral differentiation. Plant Physiol. 98: 337–342.
13. Tiburcio AF, Gendy CA, Trân Thanh Vân K (1989) Morphogenesis in tobacco subepidermal cells : putrescine as marker of root differentiation. Plant Cell Tissue and Organ Culture. 19: 43–54.
14. Trân Thanh Vân K (1973) *In vitro* and *de novo* flower, bud, root and callus differentiation from excised epidermal tissue. Nature 246: 44–45.
15. Trân Thanh Vân K (1981) Control of morphogenesis in *in vitro* cultures. Ann. Rev. Plant Physiol. 33: 291 – 311.
16. Trân Thanh Vân K ,Toubart P, Cousson A, Darvill AG, Gollin DJ, Chelf P, Albersheim P (1985) Manipulation of the morphogenetic pathways of tobacco explants by oligosaccharins. Nature 314: 615–617.
17. Trân Thanh Vân K, Richard L, Gendy C (1990) An experimental model for the analysis of plant/cell differentiation: thin cell layer. Concept, strategy, methods, records and potential. In: Durzan, Rodriguez (Eds.) NATO Biotechnology Series, pp. 215–224. Plenum Academic Press, NY.
18. Trân Thanh Vân K (1991) Molecular aspects of flowering. In: Harding, Singh, Mol JNM (Eds.) Genetics and Breeding of Ornamental Plants. Plenum Academic Press.
19. Trân Thanh Vân K (1992) *In vitro* organogenesis and somatic embryogenesis. Acta Horticulturae Propagation of Ornamental Plants 314: 27–39.
20. Trân Thanh Vân K, Gendy CA (1993) Relation between some cytological, biochemical, molecular markers and plant morphogenesis. In: Roubelakis-Angelakis, Trân Thanh Vân (Eds.) Nato Asi Series, pp. 39–54. Plenum Academic Press, NY.

Plant Tissue Culture Manual **H5**, 1–17, 1996.
© 1996 *Kluwer Academic Publishers.*

In vitro infection of *Arabidopsis* with nematodes

JOKE C. KLAP & PETER C. SIJMONS[1]
MOGEN international NV, Einsteinweg 97, 2333 CB Leiden, The Netherlands.[1] present address:
Agrotechnological Research Centre, P.O. Box 17, 6700 AA Wageningen, The Netherlands.

Introduction

Sedentary plant-parasitic nematodes establish intricate relationships with their hosts. The interaction reveals several unique features and is a prime object to study host-pathogen recognition, cell-cell communication and induction of feeding structures.

Economically the most devastating species are root-knot nematodes (e.g. *Meloidogyne incognita*) and cyst nematodes (e.g. *Heterodera schachtii*). Host ranges can be very wide and in areas of intensive agriculture, population densities may increase in 2–3 years to such levels that crop yields are severely affected. Although plant-nematode research is difficult under laboratory conditions, the recent infection protocols for *Arabidopsis* have given new momentum to this area of phytopathology [1, 2, 4, 6, 8, 9].

Nematode juveniles remain dormant in the soil until proper host roots grow in their vicinity. Diffusable root factors induce hatching and the juveniles will migrate towards young roots. They invade roots preferably behind the zone of cell elongation and move up to specific sites of the developing vascular cylinder. Here they carefully select cells for the initiation of feeding structures. This process is induced by nematode secretions [3] and will eventually lead to large, multinucleate and hypertrophic structures that provide a constant source of nourishment for the then sedentary nematode. A more detailed description of the life cycle is given in [7, 8].

The protocols described in this chapter provide complete details for the in vitro infection of *Arabidopsis* with *Heterodera schachtii*. The first section describes the maintenance of nematode starter cultures and the preparation of a sterile inoculum, followed by the infection procedure for *Arabidopsis* with freshly hatched second-stage juveniles.

Procedures

In vivo *stock of* Heterodera schachtii

Grow *Brassica oleracea* in sand and inoculate the young plants with second stage juveniles of *Heterodera schachtii*. (Other suitable hosts can be used to maintain stock cultures). The plants can be grown at standard greenhouse or growth chamber conditions although the soil temperature should be kept below 20 °C. Harvest the *Brassica* plants after 4 weeks and store the sand containing roots with cysts at 4 °C in the dark. As long as the sand remains moist, the cysts can be stored this way for several years. (Note: in some countries, quarantine regulations are necessary to grow certain species of cyst nematodes). Have *in vitro* plantlets available when you start the hatching procedure.

Collecting cysts from the sand culture

Steps in the procedure

1. Put some sand containing the roots and cysts in a beaker and add tap water. The sand will settle while the cysts, along with root debris, will float. Pour the cysts off on a 0.5 mm mesh screen that is placed in a kitchen sieve or similar device. Repeat this until you have collected all the cysts from the sand. Next, rinse the cysts on the 0.5 mm mesh screen to remove foam.
2. For easier handling, rinse the cleaned cysts from the 0.5 mm onto a 0.25 mm mesh screen and place the whole screen flat in a large Petri dish. Using fine forceps and possibly a binocular, collect only the large and clearly egg containing cysts and put them in a small sieve of 20 μm gauze (see note 1). For sterilization and hatching you can keep the cysts in this sieve (about 150 cysts per sieve).

Sterilization of the cysts and hatching of the juveniles

Steps in the procedure

1. Work from now on in a laminar flow cabinet. Put the sieve with the cysts with the aid of a sterile pair of tweezers in a 50 ml beaker and add 0.05% $HgCl_2$. Leave this for 3 minutes. Rinse the cysts 3 times in 50 ml beakers containing distilled water.
2. For hatching place the sieve in a glass funnel to which a silicon tube is attached that is closed with a clamp and place the whole unit in a beaker (Fig. 1). Add sterile 3 mM $ZnCl_2$ until the cysts are just covered. Close the top of the beaker with sterile aluminium foil and store at 23 °C in the dark.

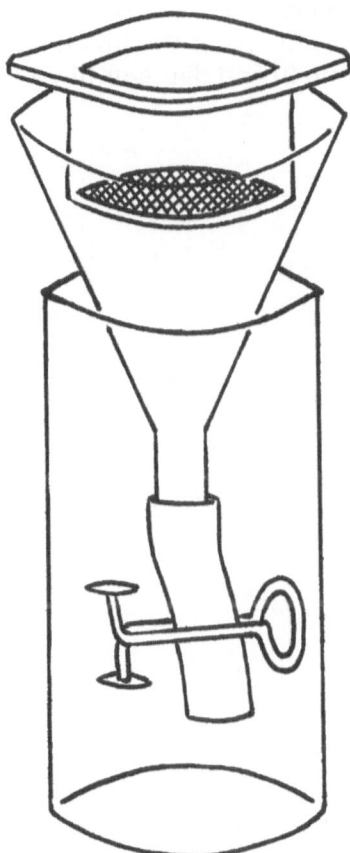

Fig. 1. Device that is used for hatching nematodes under axenic conditions. The top part is made from a syringe as described in note 1. The filter is placed in a glass funnel that is closed at the bottom with a silicon tube and a clamp. The entire unit is autoclaved before use. The hatching solution is added to just cover the cysts that are placed on the nylon filter.

3. After 3 to 4 days second stage juveniles will be visible at the bottom of the tube just above the clamp. All handling must be done in a laminar flow cabinet. Collect the juveniles by opening the clamp when the tube is above a 15 μm gauze sieve (see note 1) placed on a 25 ml beaker. Keep the juveniles on the sieve and rinse them 3 times by placing the sieve in sequential beakers filled with sterile distilled water. To surface sterilize the juveniles, place the sieve in a beaker with 0.05% $HgCl_2$ for 3 minutes and rinse again 3 times in clean beakers with sterile water. After the last rinsing the juveniles on the sieve in the beaker must be covered with water.
4. Transfer the juveniles with a sterile Pasteur's pipette from the sieve to a watch glass. Let the worms settle and decrease the volume by removing most of the water. A Pasteur's pipette that is heat-pulled into a thin capillary can be used to prevent worms being sucked up with the water. Small samples can be taken for counting the number of worms per μl. Suspend the juveniles in the watch glass in 0.5% Gelrite (see note 2) and dilute until you have the desired concentration.

In vitro *stock of* Heterodera schachtii *on* Sinapis alba *plants*

Steps in the procedure
1. Collect *Sinapis alba* seeds in a stainless steel teasieve or similar device. Soak for 2 minutes in 70% EtOH and for 10 minutes in 0.2x Teepol (0.8% active chlorite)+ 0.1% Tween 20. Rinse 3 × 10 minutes in sterile distilled water.
2. Pour 20 ml KNOP medium (0.6% agar) in 25 cm Petri dishes and let it cool to room temperature. Place two *Sinapis* seeds per dish and incubate at 23 °C (16h L, 8h D). After two weeks the plants can be inoculated with *Heterodera* second stage juveniles.
3. Inoculate the two weeks old axenic *Sinapis alba* plants with about 1000 juveniles divided in droplets of 1 µl containing about 10 juveniles per droplet. Store the inoculated *S. alba* plates at 23 °C in the dark.

Sterilisation and in vitro *culturing of* Arabidopsis *seeds*

Steps in the procedure
1. *Arabidopsis* seeds can be sterilized in several ways. We routinely use the protocol of Lluis Balcells; small seed batches (not more than a few hundred) are folded into filterpaper discs and closed with a plastified paperclip. The packages are immersed for 2 min in 70% EtOH and sterilized for 15 min in pure Teepol (4% active hypochlorite) + 0.1% Tween 20. Rinse 4 times for 10 min in sterile distilled water.
2. Pour 10 ml of solid KNOP medium (0.6% agar) in 9 cm Petri dishes (see note 3). Let the plates cool down to room temperature and place 10–20 sterilized seeds on a straight line at ca. 1/3 from the top. To synchronize germination, plates can be placed at 4 °C for 3–5 days before transfer to the growth room. During growth, the plates should be tilted at an angle of 10–30° for the roots to grow down. Two weeks after germination the roots can be inoculated.

Inoculation of Arabidopsis with sterile juveniles

Steps in the procedure
1. Collect cysts from the *Sinapis alba* roots by using a fine pair of sterilized tweezers (see note 4).
2. Hatch the juveniles without further sterilisation of the cysts. Follow for hatching the same procedure as described above.
3. As a precaution, a second sterilisation step is performed on the hatched juveniles by immersing them in 0.025% $HgCl_2$ for 3 minutes, followed by 3 rinses in sterile distilled water. Follow the same procedure as described in above.
4. Inoculate about 25 juveniles per plant suspended in droplets of 1 µl.
5. Transfer the inoculated plates back to the growth room. They can now be placed horizontal.

The infection process can be followed daily with an inverted or binocular microscope. The first signs of feeding structure development can be observed at the second or third day after inoculation. Normally, the *Heterodera* life cycle will be completed at 2–3 weeks after inoculation, when the cysts start to turn brown. Metabolic stains or detection of marker gene activities such as GUS can be done at any stage without disturbing the root system mechanically. Isolation of feeding structures from the agar-grown roots is also possible although the quantity by weight is limited. Very detailed *in situ* observations can be performed with the help of specially constructed observation chambers (Grundler, Univ. of Kiel, Germany), which allows for both root- and nematode development under high magnification.

Notes
1. Construction of the 20 µm and 15 µm gauze sieves (see also figure 1):
   - cut the top of a 25 ml syringe in a ring of about 1.5 cm in height
   - cut from 15 µm respectively 20 µm mesh gauze nylon cloth circles that are a bit larger in diameter than the syringe rings.
   - melt the gauze on the syringe by preheating the cut surface of the syringe on a hot plate of about 150 °C and quickly move the melting surface onto the nylon gauze. To improve the weld, the nylon gauze can be placed flat on a small glass plate which is heated on a second hot plate to ca. 80 °C.
   - check carefully whether the nylon gauze is completely welded onto the syringe. The juveniles can be lost during the sterilisation procedure if openings remain.
2. The gelrite is used to keep juveniles in suspension, permitting fairly even distribution during inoculation steps. The 0.5% gelrite is autoclaved for 15 minutes at 120 °C and cooled to room temperature before use.
3. It is very important not to pour more than 10 ml in standard size Petri dishes. The nematodes need root exudates to locate the roots. A larger volume will dilute this signal and leads to lower infection rates.
4. Use a binocular microscope in the laminar flow cabinet and avoid damage to the cysts as much as possible.

Solutions

Knop medium

This medium originated from W. Knop [5] and was modified for hydroponic culture. For growth of *Arabidopsis*, the medium is used at 0.2× strength. Stock solutions are stable when stored at 4 °C.

Stock solutions for Knop medium:
- KNO_3 120 gram
  $MgSO_4.7H_2O$ 19.7 gram   together in 1 liter H_2O
- $Ca(NO_3)_2.4H_2O$ 30 gram   100 ml H_2O
- KH_2PO_4 13.6 gram   100 ml H_2O
- Fe–Na EDTA   0.73 gram   100 ml H_2O
- Micronutrients/100 ml H_2O:

– H_3BO_3	286 mg
– $MnCl_2$	181 mg
– $CuCl_2.2H_2O$	5 mg
– $ZnSO_4$	3 mg
– $Na_2MoO_4.2H_2O$	7 mg
– NaCl	585 mg

To obtain 10 liter of 0.2×Knop solution, add to ca. 5 liter of water:
- KNO_3 + $MgSO_4.7H_2O$ 20 ml
- $Ca(NO_3)_2.4H_2O$ 8 ml
- KH_2PO_4 4 ml
- Fe-Na EDTA   4 ml
- Micronutrients   2 ml

Mix and fill up to 10 liter. The pH must be 6.4 (adjust with 1 N KOH; there is hardly any buffering capacity in this medium; adjust the pH carefully). Store at 4 °C in the dark.

Solid Knop medium:

Add 0.6% Daichin (or similar tissue culture grade) agar and 1% sucrose to 0.2× Knop solution. Autoclave 20 minutes at 120 °C.

Acknowledgments

We would like to thank Maria Gagiy for practical advice and Ronny Krutwagen for the drawing. This research was partially funded by Avebe, The Netherlands.

References

1. Böckenhoff A, Grundler FMW (1994) Studies on the nutrient uptake by the beet cyst nematode *Heterodera schachtii* by in situ microinjection of fluorescent probes into the feeding structures in *Arabidopsis thaliana* . Parasitology 109: 249–254.
2. Goddijn OJM, Lindsey K, Vanderlee FM, Klap JC, Sijmons PC (1993) Differential gene expression in nematode-induced feeding structures of transgenic plants harbouring promoter gusA fusion constructs. Plant J 4: 863–873.
3. Hussey RS (1989) Disease-inducing secretions of plant-parasitic nematodes. Annu Rev Phytopathol 27: 123–141.
4. Niebel A (1994) Molecular and genetic approaches to plant-nematode interactions. Thesis, State Univ. Gent, Belgium.
5. Knop W (1860) Über die Ernährung der Pflanzen durch wässerige Lösungen unter Ausschluss des Bodens. Landwirtsch Versuchsstat 2: 65–99 and 270–293.
6. Sijmons PC (1993) Plant-nematode interactions. Plant Mol Biol 23: 917–931.
7. Sijmons PC, Atkinson HJ, Wyss U (1994) Parasitic strategies of root nematodes and associated host cell responses. Annu Rev Phytopathol 32: 235–259.
8. Sijmons PC, Grundler FMW, Vonmende N, Burrows PR, Wyss U (1991) *Arabidopsis thaliana* as a new model host for plant-parasitic nematodes. Plant J 1: 245–254.
9. Wyss U, Grundler FMW (1992) *Heterodera schachtii* and *Arabidopsis thaliana*, a model host-parasite interaction. Nematologica 38: 488–493.

Index

AA medium (amino acid medium) A1/1, 2, 14, 17

ABC medium (for moss), and modifications F2/3, 38, 39

Abies spp. (first) C3/1
- transformation B13/1

Abscisic acid
- component of media A1/3–5; B13/41; C3/11, 12; C7/3; G2/18

Acalypha
- endosperm E3/1

Acer pseudoplatanus
- cell suspension cultures H3/1

Aceto-carmine
- in cytology C4/3, 5, 7

Acetolactate synthase (ALS; enzyme, gene) B9/2

Acetosyringone
- component of media B5/6; B6/5; B8/6, 11, 15; G1/5

Acetylsalicylic acid
- in protoplast culture B9/1

Achras sp.
- endosperm culture E3/2

Acridine orange
- DNA stain H3/2, 15

Actin1 gene, promoter
- to drive transgene expression B12/3, 10

Actinidia spp.
- endosperm culture E3/2, 8

Activated charcoal
- component of media A6/13; B1/9; C2/4; C3/3, 7, 11; C8/6

Adenine
- component of media C7/3, 8; E3/13

Agar
- component of media A1/6, 7; A2/1, 6, 12; A4/7, 9,11; A8/7, 9,13–15; A9/19, 23, 29; B1/5, 12; B4/17, 21; B11/3, 4, 11; C1/4–6; C3/3, 7, 8, 11, 12; C7/9; C8/5, 11, 15, 17; D5/3; D7/17; E4/5, 6; E5/17; F1/5, 6, 13; G1/5, 6; H1/18

Agarose
- protoplast embedding A1/14; A10/21, 23–25; B1/13; A11/1, 3, 4, 7; B2/4, 13; B3/7, 9, 11, 14; B4/3, 4, 13, 15, 17; B10/4–6, 10
- for callus culture A1/7; A2/1; A5/3
- for embryo culture E5/8, 13, 14

Agrobacterium rhizogenes B4/1; C8/6, 8; G1/5

- see also 'hairy roots'

Agrobacterium tumefaciens
- in plant transformation A4/2, 9; A5/1; A6/1–17; A7/1; A8/1–17; B1/1; B2/1; B4/1, 2; B5/1–9; B6/1–9; B7/1–14; B8/1–18; B10/1; B11/1–18; B12/1; B13/1; C7/5, 10; D2/1, 2; D7/1

Alkaloids
- tropane alkaloid biosynthesis in root cultures G1/1–17
- monoterpene indole alkaloids, biosynthesis G3/1–23

Amaranthus sp.
- photoautotrophic cultures H1/1

Amino acids
- components of media B10/9; E3/2, 3, 6, 13
- see also 'AA medium', 'casein hydrolysate'

Aminoglycoside phosphotransferase
- see 'neomycin phosphotransferase'

Ampicillin
- for bacterial selection C5/9
- to control culture contamination A2/16; D5/15; G1/5

Aniline blue
- fluorescent stain (pollen) D5/7, 9

Ancymidol
- gibberellin inhibitor C8/5

Annona sp.
- endosperm culture E3/2, 5

Anthocyanin
- accumulation and biosynthesis in cell cultures A3/6; A10/17; G2/1–23
- extraction and identification G2/7, 14
- enzyme assays G2/15, 16
- hormonal regulation *in vitro* G2/18, 19
- photoregulation *in vitro* G2/17, 18
- screenable gene markers D1/9; E3/8, 9

Antirrhinum majus
- anthocyanins G2/3

Apiaceae A9/1

Aphidicolin (inhibitor of DNA polymerase)
- effect on anthocyanins *in vitro* G2/19

Apple: see *Malus pumila*

Arabidopsis thaliana
- nematode infection H5/1–17
- pollen gene expression in transgenic D2/2
- protoplasts: isolation, transformation, regeneration A6/3; A7/1–20; D5/3
- regeneration and transformation A6/1–17; A8/1–17

1

3

10

- equipment A2/2; B1/13; D5/11
- filter sterilization A6/7, 12, 15; A7/4; A8/16; B1/13; B3/3, 11; B10/9; B12/17; D4/5; D5/11, 13; F1/6
- fruits (*Citrus*) C7/7
- nematodes H5/7, 8, 13
- seeds A4/5, 7; A6/5; A7/3; A8/5; A9/4–6, 23; B1/5; B2/7; B4/9; B6/3; B7/3; C3/7, 8, 11; C4/4; C8/15; E3/15; F1/6; G1/5; H5/9, 11
- tissue explants A4/9; A5/3; B1/7; B4/21; B5/5; B11/3; C1/2–5; C2/3; C6/7; C8/3, 4, 11; E3/16; E4/3; H2/5, 6; H4/7

Storage proteins
- markers of embryogenesis E2/2
- zeins E3/9

Streptomycin
- chloroplast-encoded resistance D4/3, 13; F1/13
- for bacterial selection B7/7
- to control culture contamination A2/16

Strictosidine-β-D-glucosidase
- enzyme assay G3/8, 17–19
- see also 'indole alkaloids'

Strictosidine synthase
- enzyme assay G3/8–11, 15
- see also 'indole alkaloids'

Succinic acid 2,2-methyl-hydrazide (ALAR)
- inhibitor of gibberellin synthesis C7/3

Sugarbeet: see '*Beta vulgaris*'

Sugarcane: see '*Saccharum officinarum*'

Sulfadiazine
- selective agent F2/31

Sycamore: see '*Acer pseudoplatanus*'

Synchrony
- in cell suspension cultures H3/1–31

Tabernaemontana spp.
- alkaloids G3/6

Taraxacum sp.
- endosperm E3/1

Taxillus sp.
- endosperm culture E3/3, 6, 7, 11

Taxodium spp. (bald cypress) C3/1

Taxus sp.
- transformation B13/1

Tentoxin
- chloroplast-encoded resistance D4/13

Thiamine-HCl: see 'vitamins'

Thidiazuron
- component of media C8/17

Thin cell layers (TCL)
- to study morphogenesis *in vitro* H4/1–25

Thin-layer chromatography (TLC)
- of anthocyanins G2/9–12
- of tropane alkaloids G1/9
- of indole alkaloids G3/7, 13, 15

Thula spp. (arbor vitae) C3/1

Tiglyl-CoA:pseudotropine acyl transferase
- assay G1/11, 12
- also see 'tropane alkaloids'

Tobacco: see *Nicotiana tabacum*

Tomato: see *Lycopersicon esculentum*

Tomato juice
- component of media E3/3, 5

Torenia sp.
- thin cell layers H4/2

Transient gene expression
- in protoplasts A11/1–11; B2/3, 11; B4/4, 13, 19; B10/1, 5, 6; B12/3, 11; C8/8; F2/31
- following microprojectile bombardment B13/1–46; D1/1, 5, 6, 9; D2/1–14
- see also 'chloramphenicol acetyltransferase (CAT) assay', 'β-glucuronidase (GUS) assay'

2,4,5-Trichlorophenoxyacetic acid
- component of media A1/4

Triacanthine
- component of media E3/13

Trifolium sp.
- somatic embryogenesis E2/2

Triticale
- transformation B12/2

Triticum aestivum
- chromosome isolation D7/6
- embryogenic cultures B1/3
- endosperm culture E3/3, 5, 17
- meiosis C4/7
- protoplasts B1/3
- thin cell layers H4/2, 21
- transformation B2/1; B12/1–20; D2/1

Tritordeum
- transformation B12/2

Tropane alkaloids
- biosynthesis in root cultures G1/1–17
- enzyme assays G1/11–16
- extraction and analysis G1/7–9

Tropinone reductase
- assay G1/13
- also see 'tropane alkaloids'

Tryptophan decarboxylase
- enzyme assay G3/8–13
- see also 'indole alkaloids'

Tsuga spp. (hemlock) C3/1
- transformation B13/1

Tween
  – as surfactant A2/15; A4/5, 7, 9; A8/5;
    A9/5, 6; B1/5; B6/3; B7/3; B11/3;
    B12/15; C1/2–5; C2/3; C3/7, 11; C6/8;
    C8/3, 11; E3/16

UDP-glucose pyrophosphorylase cDNA
  – B11/13, 14
UM medium
  – for cell suspension cultures D9/12
Uracil
  – yeast selective agent D1/3, 4
UV radiation
  – use as mutagen F1/2, 15; F2/17–21
Vacin and Went medium C1/1, 2, 5, 6
Vancomycin
  – antibiotic A8/3, 10, 16
Vanda sp.
  – *in vitro* culture C1/3
Verticordia spp. C8/5
Vibarabine
  – antiviral chemical C6/2
Vicia faba
  – cytological analysis C4/3
  – thin cell layers H4/2
Vicia hajastana
  – protoplasts D7/3, 5, 6, 9, 13
Virazole (ribavirin)
  – antiviral chemical C6/1, 2, 7, 8
Viruses
  – elimination A4/2; C6/1–12
  – testing C6/1–12
Viscaceae
  – endosperm culture E3/4
Vitamins
  – components of media A1/1–3, 6, 7, 9–11,
    15–17, 22; A4/2; A5/3; A6/6; A7/5;
    A8/15, 16; A9/9, 11, 15, 19, 24; B2/4;
    B3/13; B4/17; B5/2, 7, 8; B8/13–15;
    B9/10, 11; B10/9; B11/4; B12/17; B13/41,
    42; C2/4; C3/2, 12; C7/9; C9/21, 22;
    D1/7; D2/11; D5/8; D7/17; D9/12; E5/7;
    F1/6, 7, 13; F2/38; H2/7–8
  – see also 'Morel vitamins'
Vitis spp.
  – anthocyanins G2/1, 19
Vitis vinifera

  – endosperm culture E3/3
Vitrification A8/10; B8/3; C3/5, 7; C8/5
VKM medium D4/9–11

W5 salts solution A7/3, 4, 7–9; A11/3, 5, 10;
    D5/3, 6, 13; D8/2, 5; F1/5, 6
Wheat: see *Triticum aestivum*
White's medium A9/9–11; E3/4–8, 11–13, 16

X-rays
  – use as mutagen F1/2
Xylogenesis
  – in *Zinnia elegans* cell cultures: H2/1–15

Yeast: see *Saccharomyces cerevisiae*
Yeast extract
  – component of media E3/2–6; G1/6
  – see also 'Luria-Bertani medium', 'YMB
    medium'
YEB medium A6/9
YEP medium B8/11, 15
YEPD medium D1/3, 4
YMB medium G1/5, 6
Zea mays
  – embryogenic cultures B1/2, 3; B9/1
  – endosperm culture E3/3, 17
  – feeder cells E1/9
  – genes isolated D2/2, 4
  – *in vitro* fertilization E1/1–12
  – microfusion of gametes A10/3
  – pollen transfection B13/3
  – protoplasts A3/1; B1/3; B3/1; B9/1–13
  – transient gene expression D1/9
  – transgenic B1/3; B2/1; B3/1; B9/1–13;
    B12/1, 2; D1/1; D2/1
Zeatin
  – component of media A1/4; B1/13;
    B6/5–7; B12/17; C8/4, 6, 13, 17; D9/12;
    E3/2, 6, 12, 13
Zeatin riboside
  – component of media B11/4
Zeins: see 'Storage proteins'
Zephiran: see 'benzalkonium chloride'
Zinnia elegans
  – mesophyll culture system H2/1–15
Zygotic embryos
  – *in vitro* culture E5/1–19

15